Preston Albert Lambert

Differential and Integral Calculus for Technical Schools and Colleges

Preston Albert Lambert

Differential and Integral Calculus for Technical Schools and Colleges

ISBN/EAN: 9783744645799

Printed in Europe, USA, Canada, Australia, Japan

Cover: Foto ©berggeist007 / pixelio.de

More available books at **www.hansebooks.com**

Digitized by the Internet Archive
in 2008 with funding from
Microsoft Corporation

http://www.archive.org/details/differentialinte00lambrich

DIFFERENTIAL AND INTEGRAL CALCULUS

DIFFERENTIAL AND INTEGRAL CALCULUS

FOR

TECHNICAL SCHOOLS AND COLLEGES

BY

P. A. LAMBERT, M.A.

ASSISTANT PROFESSOR OF MATHEMATICS, LEHIGH UNIVERSITY

New York
THE MACMILLAN COMPANY
LONDON: MACMILLAN & CO., Ltd.
1898

All rights reserved

COPYRIGHT, 1898,
BY THE MACMILLAN COMPANY.

Norwood Press
J. S. Cushing & Co. — Berwick & Smith
Norwood Mass. U.S.A.

PREFACE

This text-book on the Differential and Integral Calculus is intended for students who have a working knowledge of Elementary Geometry, Algebra, Trigonometry, and Analytic Geometry.

The object of the text-book is threefold:

By a logical presentation of principles to inspire confidence in the methods of infinitesimal analysis.

By numerous problems to aid in acquiring facility in applying these methods.

By applications to problems in Physics and Engineering, and other branches of Mathematics, to show the practical value of the Calculus.

The division of the subject-matter according to classes of functions, makes it possible to introduce these applications from the start, and thereby arouse the interest of the student.

The simultaneous treatment of differentiation and integration, and the use of trigonometric substitution to simplify integration, economize the time and effort of the student.

P. A. LAMBERT.

TABLE OF CONTENTS

CHAPTER I

On Functions

ARTICLE   PAGE
1. Definition of a Function 1
2. The Indefinitely Large and the Indefinitely Small . . . 2
3. Limits 3
4. Corresponding Differences of Function and Variable . . 5
5. Classification of Functions 6

CHAPTER II

The Limit of the Ratio and the Limit of the Sum

6. Direction of a Curve 9
7. Velocity 12
8. Rate of Change 14
9. The Limit of the Sum 15
10. General Theory of Limits 17
11. Continuity 18

CHAPTER III

Differentiation and Integration of Algebraic Functions

12. Differentiation 21
13. Integration 29
14. Definite Integrals 34
15. Evaluation of the Limit of the Sum 36
16. Infinitesimals and Differentials 37

CHAPTER IV

Applications of Algebraic Differentiation and Integration

17. Tangents and Normals 43
18. Length of a Plane Curve 45

CONTENTS

ARTICLE		PAGE
19.	Area of a Plane Surface	48
20.	Area of a Surface of Revolution	50
21.	Volume of a Solid of Revolution	52
22.	Solids generated by the Motion of a Plane Figure	53

CHAPTER V

SUCCESSIVE ALGEBRAIC DIFFERENTIATION AND INTEGRATION

23.	The Second Derivative	57
24.	Maxima and Minima	63
25.	Derivatives of Higher Orders	74
26.	Evaluation of the Indeterminate Form $\frac{0}{0}$	75

CHAPTER VI

PARTIAL DIFFERENTIATION AND INTEGRATION OF ALGEBRAIC FUNCTIONS

27.	Partial Differentiation	78
28.	Partial Integration	82
29.	Differentiation of Implicit Functions	83
30.	Successive Partial Differentiation and Integration	86
31.	Area of Any Curved Surface	89
32.	Volume of Any Solid	91
33.	Total Differentials	92
34.	Differentiation of Indirect Functions	96
35.	Envelopes	97

CHAPTER VII

CIRCULAR AND INVERSE CIRCULAR FUNCTIONS

36.	Differentiation of Circular Functions	100
37.	Evaluation of the Forms $\infty \cdot 0$, $\infty - \infty$, $\frac{\infty}{\infty}$	105
38.	Integration of Circular Functions	107
39.	Integration by Trigonometric Substitution	111
40.	Polar Curves	113
41.	Volume of a Solid by Polar Space Coordinates	117
42.	Differentiation of Inverse Circular Functions	118
43.	Integration by Inverse Circular Functions	120
44.	Radius of Curvature	123

CHAPTER VIII

Logarithmic and Exponential Functions

ARTICLE		PAGE
45.	The Limit of $\left(1 + \dfrac{1}{z}\right)^z$ when Limit $z = \infty$	128
46.	Differentiation of Logarithmic Functions	131
47.	Integration by Logarithmic Functions	133
48.	Integration by Partial Fractions	135
49.	Integration by Parts	139
50.	Integration by Rationalization	142
51.	Evaluation of Forms 1^∞, ∞^0, 0^0	144
52.	Differentiation of Exponential Functions	145
53.	Integration of Exponential Functions	147
54.	The Hyperbolic Functions	148
55.	The Definite Integral $\int_{-\infty}^{+\infty} e^{-x^2} \cdot dx$	150
56.	Differentiation of a Definite Integral	151
57.	Mean Value	152

CHAPTER IX

Center of Mass and Moment of Inertia

58.	Center of Mass	154
59.	Center of Mass of Lines	155
60.	Center of Mass of Surfaces	157
61.	Center of Mass of Solids	161
62.	Theorems of Pappus	162
63.	Moment of Inertia	165
64.	Moment of Inertia of Lines and Surfaces	167
65.	Moment of Inertia of Solids	170

CHAPTER X

Expansions

66.	Convergent Power Series	173
67.	Taylor's and Maclaurin's Series	175
68.	Euler's Formulas for Sine and Cosine	184
69.	Differentiation and Integration of Power Series	186
70.	Expansion of $u_1 \equiv f(x + h, y + k)$	192

CHAPTER XI

APPLICATIONS OF TAYLOR'S SERIES

ARTICLE		PAGE
71.	Maxima and Minima by Expansion	194
72.	Contact of Plane Curves	198
73.	Singular Points of Plane Curves	201

CHAPTER XII

ORDINARY DIFFERENTIAL EQUATIONS OF FIRST ORDER

74.	Formation of Differential Equations	205
75.	Solution of First Order Differential Equations of First Degree	209
76.	Equations of First Order and Higher Degrees	218
77.	Ordinary Equations in Three Variables	222

CHAPTER XIII

ORDINARY DIFFERENTIAL EQUATIONS OF HIGHER ORDER

78.	Equations of Higher Order and First Degree	225
79.	Symbolic Integration	230
80.	Symbolic Solution of Linear Equations	234
81.	Systems of Simultaneous Differential Equations	237

CHAPTER XIV

PARTIAL DIFFERENTIAL EQUATIONS

82.	Formation of Partial Differential Equations	239
83.	Partial Differential Equations of First Order	240
84.	Linear Equations of Higher Order	244

DIFFERENTIAL AND INTEGRAL CALCULUS*

CHAPTER I

ON FUNCTIONS

Art. 1. — Definition of a Function

A constant is a quantity which retains the same value throughout a discussion.

A variable is a quantity which has different successive values in the same discussion.

If two variables are so related that to every value of the first there correspond one or more determinate values of the second, the second is called a function of the first.

Denote the variables by x and y, and let the relation between them be expressed by the equation $y = mx + n$, where m and n are constants. If in a particular discussion $m = 3$ and $n = 5$, the equation becomes $y = 3x + 5$. Arbitrary values may be assigned to x, and the corresponding values of y calculated. If in another discussion $m = -2$ and $n = 6$, the equa-

---

* The Differential and Integral Calculus was invented independently by Newton and Leibnitz, Newton antedating Leibnitz by several years. Leibnitz published his method in the "Acta Eruditorum" of Leipzig, in 1684; Newton published his method in his "Natural Philosophy," in 1687.

tion becomes $y = -2x + 6$. Again, arbitrary values may be assigned to x, and the corresponding values of y calculated.

The variable x, to which arbitrary values are assigned, is called the independent variable. The variable y, whose value depends on x, is called the dependent variable or function.

In the equation $y = 3x + 5$, if x increases, y increases; if x decreases, y decreases. This fact is expressed by calling y an increasing function of x.

In the equation $y = \dfrac{3}{x+5}$, if x increases, y decreases; if x decreases, y increases. This fact is expressed by calling y a decreasing function of x.

Art. 2. — The Indefinitely Large and Indefinitely Small

The term of the geometric progression $1, 2, 2^2, 2^3, 2^4, 2^5, \cdots$ continually increases, and becomes larger than any number that can be assigned when the progression is extended sufficiently far.

The term of the geometric progression $1, \dfrac{1}{2}, \dfrac{1}{2^2}, \dfrac{1}{2^3}, \dfrac{1}{2^4}, \dfrac{1}{2^5} \cdots$ continually decreases and becomes smaller than any number that can be assigned when the progression is extended sufficiently far.

A variable quantity whose numerical value continually increases and becomes larger than any quantity that can be assigned is said to become indefinitely large.

A variable quantity whose numerical value continually decreases and becomes smaller than any quantity that can be assigned, is said to become indefinitely small.

If $y = \dfrac{3}{x+5}$, and x is positive and becomes indefinitely large, the corresponding value of y is positive, and becomes indefinitely small. If x is negative, and starting from zero continually approaches -5 in such a manner that the difference between the value assigned to x and -5 becomes indefinitely small, the corresponding value of y is positive and becomes indefinitely large. If x continues to decrease beyond -5, y becomes negative and its numerical value decreases, becoming indefinitely small when the numerical value of x becomes indefinitely large.

All quantities which lie between those which are indefinitely small and those which are indefinitely large are called finite.

If ϵ denotes an indefinitely small quantity, and n is finite, the product $n \cdot \epsilon$ must also be indefinitely small. For, if $n \cdot \epsilon = m$, and m is finite, $\epsilon = \dfrac{m}{n}$, the ratio of two finite quantities. Hence ϵ would be a quantity whose value can be assigned, which is contrary to the hypothesis.

Since $y(x+5) = 3$ is always true if $y = \dfrac{3}{x+5}$, and when x becomes indefinitely large, y becomes indefinitely small, it follows that the product of an indefinitely small quantity and an indefinitely large quantity may be finite.

The sum of a finite number of indefinitely small quantities is indefinitely small. For if ϵ is the largest of the n indefinitely small quantities $\epsilon_1, \epsilon_2, \epsilon_3, \epsilon_4, \cdots, \epsilon_n$, their sum

$$\epsilon_1 + \epsilon_2 + \epsilon_3 + \epsilon_4 + \cdots + \epsilon_n$$

cannot be greater than $n \cdot \epsilon$, which is indefinitely small when n is finite.

Art. 3. — Limits

If one quantity continually approaches a second quantity in such a manner that the difference between the two becomes indefinitely small, the second quantity is called the limit of the first.

For example, if $y = 5 + \dfrac{3}{x}$ when x becomes indefinitely large, $\dfrac{3}{x}$ becomes indefinitely small, and by the definition 5 is the limit of y.

The limit of a quantity which becomes indefinitely small is zero.

The limit of a quantity which becomes indefinitely large is called infinity, and is denoted by the symbol ∞.

This conception of a limit is used in elementary geometry when the circumference of a circle is proved to be the limit of the perimeter of the inscribed regular polygon when the number of sides becomes indefinitely large; when the area of a circle is proved to be the limit of the area of the inscribed regular polygon when the number of sides becomes indefinitely large; when the volume of a triangular pyramid is proved to be the limit of the sum of the volumes of inscribed triangular prisms of equal altitude and with bases parallel to the base of the pyramid when the number of prisms becomes indefinitely large.

In elementary algebra, the sum of n terms of the geometric progression $a,\ a \cdot r,\ a \cdot r^2,\ a \cdot r^3,\ a \cdot r^4,\ \cdots$ is proved to be

$$s_n = \frac{a - a \cdot r^n}{1 - r} = \frac{a}{1 - r} - \frac{a \cdot r^n}{1 - r}.$$

If r is numerically less than unity and a is finite, $\dfrac{a \cdot r^n}{1 - r}$ becomes indefinitely small when n becomes indefinitely large. Hence when n becomes indefinitely large, the limit of s_n is $\dfrac{a}{1 - r}$.

If r is positive, the successive values of s_n as n increases all lie on the same side of the limit; if r is negative, the successive values of s_n oscillate from one side of the limit to the other. For example, the limit of the sum of an indefinitely large number of terms of the geometric progression $1, -\tfrac{1}{2}, \tfrac{1}{4}, -\tfrac{1}{8}, \tfrac{1}{16}, -\tfrac{1}{32}, \tfrac{1}{64}, -\tfrac{1}{128}, \tfrac{1}{256}, -\tfrac{1}{512}, +\cdots$, is $\tfrac{2}{3}$. The first ten successive approximations are $s_1 = 1$, $s_2 = \tfrac{1}{2}$, $s_3 = \tfrac{3}{4}$, $s_4 = \tfrac{5}{8}$, $s_5 = \tfrac{11}{16}$, $s_6 = \tfrac{21}{32}$, $s_7 = \tfrac{43}{64}$, $s_8 = \tfrac{85}{128}$, $s_9 = \tfrac{171}{256}$, $s_{10} = \tfrac{341}{512}$. The approximation s_n is larger than the limit when n is odd, smaller than the limit when n is even.

If $y = \dfrac{x^2 - 1}{x - 1}$, y has a determinate value for every value of x, except for $x = 1$. When $x = 1$, y takes the indeterminate form $\dfrac{0}{0}$, which may have any value whatever. The true value of y when $x = 1$ is defined as the limit of the values of y corresponding to values of x whose limit is 1. For example,

when $x = 1.1,\ 1.01,\ 1.001,\ 1.0001,\ 1.00001,\ 1.000001,\ \cdots$,

$y = 2.1,\ 2.01,\ 2.001,\ 2.0001,\ 2.00001,\ 2.000001,\ \cdots$.

Hence 2 is the true value of y when $x = 1$, and for all values of x, including $x = 1$, $y = \dfrac{x^2 - 1}{x - 1} \equiv x + 1$.

ART. 4. — CORRESPONDING DIFFERENCES OF FUNCTION AND VARIABLE

If $y = x^2$, when $x = -4,\ -3,\ -2,\ -1,\ 0,\ 1,\ 2,\ 3,\ 4$,
$y = 16,\ 9,\ 4,\ 1,\ 0,\ 1,\ 4,\ 9,\ 16$.

Starting from $x = 0$, a difference of $+1$ in the value of x causes a difference of $+1$ in the value of y; starting from

$x = +1$, a difference of $+1$ in the value of x causes a difference of $+3$ in the value of y; starting from $x = -3$, a difference of $+1$ in the value of x causes a difference of -5 in the value of y. Observe that the same change in the value of the variable in different parts of the function causes different changes in the value of the function.

In general, if $y = x^2$, and starting from any value of x the corresponding differences of y and x are denoted by Δy and Δx, so that $x + \Delta x$ and $y + \Delta y$ must satisfy the equation $y = x^2$, by subtracting the equations $y + \Delta y = (x + \Delta x)^2$ and $y = x^2$, there results $\Delta y = 2x \cdot \Delta x + (\Delta x)^2$. The difference in the value of y corresponding to a difference of Δx in the value of x is seen to depend on x and on Δx.

Art. 5. — Classification of Functions

A function is called algebraic if the relation between function and variable can be expressed by means of a finite number of the fundamental operations of algebra, addition, subtraction, multiplication, division, and involution and evolution with constant indices.

For example, $y = x^3 - 7x + 7$ explicitly defines y as an integral, rational, one-valued, algebraic function of x. The equation $x^2 + y^2 = 9$ implicitly defines y as a two-valued algebraic function of x. The relation $y = \dfrac{x^2 - 1}{1 + x^{\frac{1}{2}}}$ defines y as a fractional, irrational, one-valued function of x.

All functions not algebraic are called transcendental. The expression of transcendental functions by means of the fundamental operations of algebra is possible only in the form of the sum of an indefinitely large number of terms, or in the form of the product of an indefinitely large number of factors.

ON FUNCTIONS

The elementary transcendental functions are:

The exponential function $y = a^x$ and its inverse the logarithmic function $x = \log_a y$.

The trigonometric or circular functions $y = \sin x$, $y = \tan x$, $y = \sec x$, together with their complementary functions; and the inverse functions $x = \sin^{-1} y$, $x = \tan^{-1} y$, $x = \sec^{-1} y$.

In general, the fact that y is an explicit function of x, without specifying the nature of the function, is denoted by writing $y = f(x)$, or $y = F(x)$, or $y = \phi(x)$; the fact that y is an implicit function of x is denoted by writing $f(x, y) = 0$, or $F(x, y) = 0$, or $\phi(x, y) = 0$.

PROBLEMS

1. Determine the values of the function $y = 3x^2 - 5$ corresponding to $x = 0, 1, 2, 3, 4, 5$.

2. Find the true value of $y = \dfrac{x^2 - 4}{x - 2}$ when $x = 2$.

3. Find the true value of $y = \dfrac{x^3 - 1}{x - 1}$ when $x = 1$.

4. Starting from $x = 3$, calculate the difference in y corresponding to a difference of 2 in the value of x if
$$y = x^2 - 5x + 12.$$

5. Starting from $x = 2$, calculate the difference in y corresponding to a difference of -2 in the value of x if $y = 7x - 3x^2$.

6. Find the limit of the sum of an indefinitely large number of terms of the series $1 + \frac{1}{3} + \frac{1}{9} + \frac{1}{27} + \frac{1}{81} + \cdots$.

7. Find the limit of the sum of an indefinitely large number of terms of the series $\frac{3}{10} + \frac{3}{100} + \frac{3}{1000} + \frac{3}{10000} + \cdots$.

8. Show that any difference in the abscissa of the straight line whose equation is $y = mx + n$ causes a difference m times as large in the ordinate.

9. Show that the ordinate of the straight line whose equation is $y - y_0 = m(x - x_0)$ changes m times as fast as the abscissa.

10. If $y = 3x^2 - 7x$, determine the change Δy in the value of y corresponding to a change of Δx in the value of x.

11. Compare the values of Δy corresponding to $\Delta x = 1$, starting from $x = 0, 1, 2, 3$, if $y = x^2 - 3x + 10$.

12. If $y = ax^2 + bx + c$, where a, b, and c are constants and when $x = x_0$, $y = y_0$, determine the change Δy in the value of y corresponding to a change of Δx in the value of x, starting from $x = x_0$.

CHAPTER II

THE LIMIT OF THE RATIO AND THE LIMIT OF THE SUM

Art. 6. — Direction of a Curve

Let (x_0, y_0) be any point of the curve whose equation is $y = x^2$. Let $(x_0 + \Delta x, y_0 + \Delta y)$ be any other point of the curve, Δx and Δy representing corresponding differences in abscissa and ordinate. The ratio $\frac{\Delta y}{\Delta x}$ is the slope of the secant line through (x_0, y_0) and $(x_0 + \Delta x, y_0 + \Delta y)$, that is, the tangent of the angle of inclination of the secant to the X-axis. The equation of the secant is

$$y - y_0 = \frac{\Delta y}{\Delta x}(x - x_0).$$

This is true whatever may be the magnitude of the corresponding differences Δx and Δy.

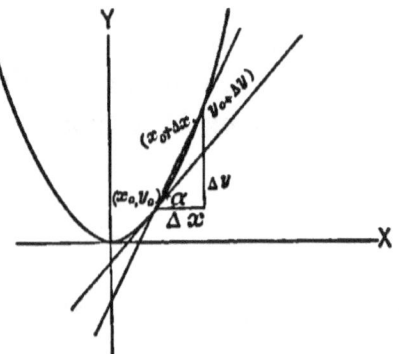

Fig. 1.

Now the tangent to a curve is defined as the limiting position of the secant whose two points of intersection are made to continually approach each other. If the point $(x_0 + \Delta x, y_0 + \Delta y)$ continually approaches the point (x_0, y_0), the limit of Δx is zero, and the corresponding limit of the ratio $\frac{\Delta y}{\Delta x}$

is the slope of the tangent to the curve at (x_0, y_0). This limit of the ratio is to be determined.

By hypothesis, $y_0 + \Delta y = (x_0 + \Delta x)^2$, $y_0 = x_0^2$; whence

$$\frac{\Delta y}{\Delta x} = \frac{(x_0 + \Delta x)^2 - x_0^2}{\Delta x}.$$

The limit of this ratio, when the limit of Δx is zero, cannot be found by placing $\Delta x = 0$ in this value of the ratio. For this makes the ratio take the indeterminate form $\frac{0}{0}$ as it ought to, for the two points are made coincident, and through one point an infinite number of straight lines may be drawn. By performing the operations indicated in the numerator of the value of the ratio $\frac{\Delta y}{\Delta x}$ and then dividing out the factor Δx common to numerator and denominator, there results

$$\frac{\Delta y}{\Delta x} = 2x_0 + \Delta x_0.$$

If now Δx becomes indefinitely small, that is, if the limit of Δx is zero, the limit of the ratio $\frac{\Delta y}{\Delta x}$ is $2x_0$.

Denoting by α the angle of inclination to the X-axis of the tangent to $y = x^2$ at (x_0, y_0), $\tan \alpha = 2x_0$ and the equation of the tangent is $y - y_0 = 2x_0(x - x_0)$. At the point $(2, 4)$, $\tan \alpha = 4$ and $\alpha = 75° 58'$. The tangent makes an angle of $45°$ with the X-axis if $2x_0 = \tan 45° = 1$. Solving the equations $2x_0 = 1$ and $y_0 = x_0^2$, the point of tangency is found to be $x_0 = \frac{1}{2}$, $y_0 = \frac{1}{4}$.

If the curve whose equation is $y = x^2$ is generated by the continuous motion of a point, when the generating point passes the point (x_0, y_0) of the curve it tends at that instant to move along the tangent $y - y_0 = 2x_0(x - x_0)$ at the point (x_0, y_0).

Hence the direction of the tangent to the curve at any point is called the direction of the curve at that point.

If a point moves along the straight line $y - y_0 = 2x_0(x - x_0)$, the ordinate changes $2x_0$ times as fast as the abscissa. When $2x_0$ is positive, the ordinate is an increasing function of the abscissa; when $2x_0$ is negative, the ordinate is a decreasing function of the abscissa; when $2x_0 = 0$, the line is parallel to the X-axis and a change in the abscissa causes no change in the ordinate.

Hence, when the point generating the curve $y = x^2$ passes the point (x_0, y_0), the ordinate of the curve is at that instant changing value $2x_0$ times as fast as the abscissa changes. At the point $(\frac{1}{2}, \frac{1}{4})$ ordinate and abscissa are changing value at the same rate; at the point $(2, 4)$ the ordinate is increasing 4 times as fast as the abscissa increases; at the point $(-2, 4)$ the ordinate decreases 4 times as fast as the abscissa increases.

By precisely the same analysis it is proved that the slope of the tangent to the curve whose equation is $y = f(x)$ at any point (x, y) of the curve is $\tan \alpha = \text{limit}\,\frac{\Delta y}{\Delta x} = \text{limit}\,\frac{f(x + \Delta x) - f(x)}{\Delta x}$, when the limit of Δx is zero, and that the limit of this ratio measures the rate of change of ordinate and abscissa at (x, y). It is essential to remember that in the calculation of this limit Δx must start from some finite value and then be made to approach the limit zero.*

PROBLEMS

1. Find the slope of the tangent to $y = x^3 - 3x$ at $x = 5$.

2. Find the direction in which the point generating the graph of $y = 3x^2 - x$ tends to move, when $x = 1$.

---

* The tangent problem prepared the way for the invention of the Differential Calculus, in the seventeenth century.

3. Find the rate of change of ordinate and abscissa of $y = 3x^2 - x$ at $x = 1$.

4. Find at what point of the curve whose equation is $y = 4x^2$ the tangent makes with the X-axis an angle of $45°$.

5. Find the equation of the tangent to $y = 2x^2 - 5x$ at $x = 3$.

6. Find where the ordinate of $y = 3x - 4x^2$ decreases 5 times as fast as x increases.

ART. 7. — VELOCITY

Suppose a locomotive to start at station A, to pass station B distant s_0 miles from A after t_0 hours, and station C distant s miles from A after t hours. The average velocity per hour

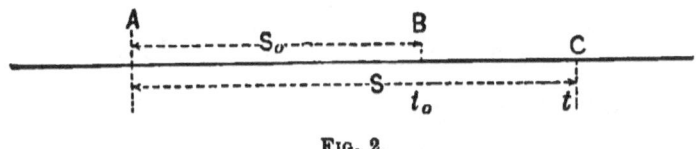

FIG. 2.

from B to C, that is, the uniform number of miles per hour the locomotive must run from B to C to cover the distance $s - s_0$ miles in $t - t_0$ hours, is $\frac{s - s_0}{t - t_0}$. Calling the difference of distance Δs and the difference of time Δt, the average velocity is $\frac{\Delta s}{\Delta t}$. The equal ratios $\frac{\Delta s}{\Delta t} = \frac{s - s_0}{t - t_0}$ determine the average velocity of the locomotive during the interval of time $\Delta t = t - t_0$, whatever may be the magnitude of this interval of time.

Now if station C is taken nearer and nearer station B, the interval of time Δt becomes indefinitely small and has zero for limit. The average velocity from B to C continually approaches the actual velocity at B, since the interval of time during which a change of velocity might take place continually decreases. Hence the limit of the ratio $\frac{\Delta s}{\Delta t}$ when the limit of Δt is zero is the actual velocity of the locomotive at B.

This analysis shows that if the relation between distance s and time t of the motion of a body is expressed by the equation $s = f(t)$, and the velocity at any time t is denoted by v,

$v = \text{limit } \frac{f(t + \Delta t) - f(t)}{\Delta t}$ when the limit of Δt is zero.

For example, in the case of a freely falling body, starting from rest, $s = 16.08\, t^2$, where s is distance measured in feet, and t is time measured in seconds. Here

$$v = \text{limit } \frac{16.08\,(t + \Delta t)^2 - 16.08\, t^2}{\Delta t}$$

$$= \text{limit } 16.08\,(2t + \Delta t) = 32.16\, t,$$

when the limit of Δt is zero. Hence the velocity at the end of the third second is 96.48 feet per second.

Velocity is seen to be the rate of change of distance per unit of time.

PROBLEMS

1. If $s = \frac{1}{2} gt^2$, where g is a constant, determine the velocity at time t.

2. If $s = 10t + 16.08\, t^2$, calculate the velocity at time t.

3. If $s = ut - \frac{1}{2} gt^2$, where u and g are constants, determine the velocity at time t.

4. If $s = ut + \frac{1}{2} gt^2$, where u and g are constants, determine the velocity at time t.

Art. 8. — Rate of Change

In the function $y = x^2 - 5x$ let x_0, y_0 and $x_0 + \Delta x, y_0 + \Delta y$ be two sets of corresponding values of variable and function. That is, starting from x_0, y_0 to a change of Δx in the value of the variable there corresponds a change of Δy in the value of the function. The ratio of the corresponding changes of function and variable $\frac{\Delta y}{\Delta x}$ determines the average rate of change of the function throughout the interval Δx; that is, the uniform change of the function for change of the variable by unity which in the interval Δx causes a change of Δy in the value of the function. This is true for all values of the interval Δx.

Now, if Δx becomes smaller and smaller, the ratio $\frac{\Delta y}{\Delta x}$ continually approaches the actual rate of change of the function at x_0, since the interval Δx during which the rate of change might vary continually decreases. Hence the actual rate of change of the function $y = x^2 - 5x$ at x_0, y_0 is

$$\text{limit } \frac{\Delta y}{\Delta x} = \text{limit } \frac{(x_0 + \Delta x)^2 - 5(x_0 + \Delta x) - (x_0^2 - 5x_0)}{\Delta x}$$
$$= \text{limit } (2x_0 - 5 + \Delta x) = 2x_0 - 5,$$

when the limit of Δx is zero.

The function $y = x^2 - 5x$ increases 3 times as fast as x increases when $2x_0 - 5 = 3$, that is, when $x_0 = 4$; the function decreases 5 times as fast as x increases when $2x_0 - 5 = -5$, that is, when $x_0 = 0$; the function is stationary, that is, it is neither increasing nor decreasing, when $2x_0 - 5 = 0$.

This analysis shows that for any value of x the limit of $\frac{f(x + \Delta x) - f(x)}{\Delta x}$ when the limit of Δx is zero, measures the rate of change of $f(x)$ for that value of x. If this limit is positive, $f(x)$ is an increasing function of x; if this limit is negative, $f(x)$ is a decreasing function of x; if this limit is zero, $f(x)$ is stationary.

The calculation of the limit of the ratio $\dfrac{f(x+\Delta x)-f(x)}{\Delta x}$ when limit $\Delta x = 0$ for all functions $f(x)$ is the fundamental problem of the Differential Calculus.

PROBLEMS

1. Calculate the rate of change of $4x + 7$.
2. Find the rate of change of $y = x^3 + 3x$.
3. Calculate the rate of change of $3x - x^2$ at $x = 1$.
4. Find where the function $x^2 - 2x$ increases twice as fast as x increases.
5. Find where the function $x^2 - 2x$ decreases twice as fast as x increases.
6. Find where the function $x^2 - 2x$ is stationary.

Art. 9.— The Limit of the Sum

Let $y = f(x)$ be the equation of the given curve. Denote by A the area of the surface bounded by the curve, the X-axis, and the lines $x = a$, $x = b$. Divide the portion of the X-axis from $x = a$ to $x = b$ into any number, say 5, of equal parts, and call each part Δx. Constructing rectangles on each Δx, as indicated in the figure, and denoting by A' the

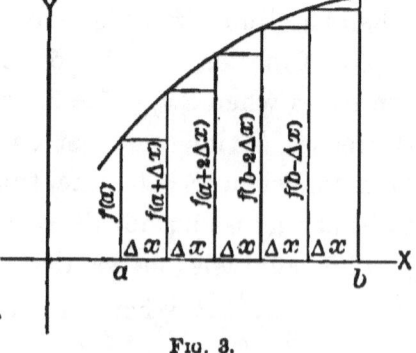

Fig. 3.

sum of the areas of the rectangles,

$$A' = f(a) \cdot \Delta x + f(a + \Delta x) \cdot \Delta x + f(a + 2\Delta x) \cdot \Delta x$$
$$+ f(b - 2\Delta x) \cdot \Delta x + f(b - \Delta x) \cdot \Delta x,$$

which may be written $A' = \sum_{x=a}^{x=b} f(x) \cdot \Delta x$. Now, as the number of equal parts into which $b - a$ is divided is indefinitely increased, Δx becomes indefinitely small, and A' continually approaches A. Hence $A = $ limit of $\sum_{x=a}^{x=b} f(x) \cdot \Delta x$ when the limit of Δx is zero. The calculation of this limit of the sum is one of the fundamental problems of the Integral Calculus.*

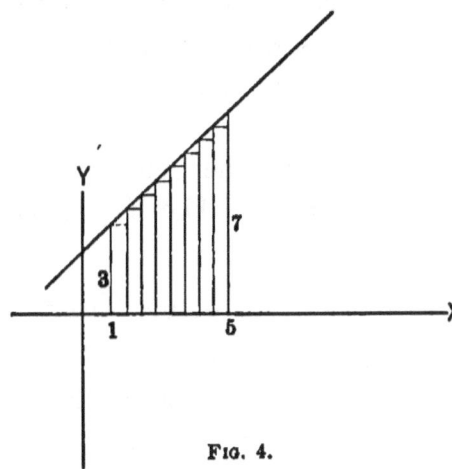

Fig. 4.

In some simple problems the limit of the sum may be calculated by means of the formula for the sum of n terms of an arithmetic progression

$$a + (a + d) + (a + 2d) + \cdots$$
$$+ (l - 2d) + (l - d) + l,$$

namely, $s = (a + l)\dfrac{n}{2}$.

For example, let it be required to calculate the area of the surface bounded by the straight lines $y = x + 2$, $x = 1$, $x = 5$, and the X-axis. Dividing the distance from $x = 1$ to $x = 5$ into n equal parts and calling each part Δx, $\Delta x = \dfrac{4}{n}$, and

---

* Historically the calculation of the areas of surfaces bounded by curved lines led to the invention of the Integral Calculus.

$$A = \text{limit} \sum_{x=1}^{x=4} f(x)\,\Delta x$$

$$= \text{limit} \left\{ 3 + \left(3 + \frac{4}{n}\right) + \left(3 + \frac{8}{n}\right) + \left(3 + \frac{12}{n}\right) + \cdots \right.$$
$$\left. + \left(3 + [n-1]\frac{4}{n}\right) \right\} \frac{4}{n}$$

$$= \text{limit} \left\{ 6 + (n-1)\frac{4}{n} \right\} \frac{n}{2} \cdot \frac{4}{n} = \text{limit}\, 2\left(10 - \frac{4}{n}\right) = 20$$

when n is indefinitely increased. This agrees with the result obtained by elementary geometry.

Art. 10. — General Theory of Limits

Let u denote a function of x whose limit is U when the limit of x is a. This relation may be denoted by the equation $u_{(x=a\pm\delta)} = U \pm \epsilon$, where ϵ must become indefinitely small when δ becomes indefinitely small.

Limit of the sum. — Suppose that when the limit of x is a, limit $u_1 = U_1$, limit $u_2 = U_2$, limit $u_3 = U_3$. The hypothesis is equivalent to

$$u_{1(x=a\pm\delta)} = U_1 \pm \epsilon_1, \quad u_{2(x=a\pm\delta)} = U_2 \pm \epsilon_2, \quad u_{3(x=a\pm\delta)} = U_3 \pm \epsilon_3,$$

whence $(u_1 + u_2 - u_3)_{(x=a\pm\delta)} = U_1 + U_2 - U_3 \pm \epsilon_1 \pm \epsilon_2 \mp \epsilon_3$. Since when δ becomes indefinitely small, $\epsilon_1, \epsilon_2, \epsilon_3$ each become indefinitely small, $\pm \epsilon_1 \pm \epsilon_2 \mp \epsilon_3$ for all combinations of signs also becomes indefinitely small. Hence when the limit of x is a, limit $(u_1 + u_2 - u_3) = U_1 + U_2 - U_3 =$ limit $u_1 +$ limit $u_2 -$ limit u_3. That is, the limit of the algebraic sum of a finite number of quantities is the like algebraic sum of their limits.

Limit of the product. — If, when the limit of x is a, limit $u_1 = U_1$ and limit $u_2 = U_2$, $u_{1(x=a\pm\delta)} = U_1 \pm \epsilon_1$ and $u_{2(x=a\pm\delta)} = U_2 \pm \epsilon_2$. By multiplication,

$$(u_1 \cdot u_2)_{(x=a\pm\delta)} = U_1 \cdot U_2 \pm U_2 \cdot \epsilon_1 \pm U_1 \cdot \epsilon_2$$

c

Hence if U_1 and U_2 are finite, when the limit of x is a, limit $(u_1 \cdot u_2) = U_1 \cdot U_2 = \text{limit } u_1 \cdot \text{limit } u_2$. That is, the limit of the product is the product of the limits.

Limit of the quotient. — If, when the limit of x is a, limit $u_1 = U_1$ and limit $u_2 = U_2$, $u_{1\,(x=a\pm\delta)} = U_1 \pm \epsilon_1$, $u_{2\,(x=a\pm\delta)} = U_2 \pm \epsilon_2$. By division,

$$\left(\frac{u_1}{u_2}\right)_{(x=a\pm\delta)} = \frac{U_1 \pm \epsilon_1}{U_2 \pm \epsilon_2} = \frac{U_1}{U_2} + \frac{U_1 \pm \epsilon_1}{U_2 \pm \epsilon_2} - \frac{U_1}{U_2}$$

$$= \frac{U_1}{U_2} + \frac{\pm U_2 \cdot \epsilon_2 \mp U_1 \cdot \epsilon_2}{U_2(U_2 \pm \epsilon_2)}.$$

Hence if U_1 and U_2 are finite, when the limit of x is a, $\text{limit } \dfrac{u_1}{u_2} = \dfrac{U_1}{U_2} = \dfrac{\text{limit } u_1}{\text{limit } u_2}$. That is, the limit of the quotient is the quotient of the limits.*

Art. 11. — Continuity

The function $y = f(x)$ is said to be continuous at $x = x_0$ if the limit of the difference $f(x_0 \pm \Delta x) - f(x_0)$ is zero when the limit of Δx is zero. This definition may also be written $[f(x_0 + \Delta x) - f(x_0)]_{\Delta x = \pm \delta} = \pm \epsilon$, where ϵ must become indefinitely small when δ becomes indefinitely small. The function is said to be discontinuous at $x = x_0$ if ϵ does not become indefinitely small when δ becomes indefinitely small.

For example, if the curve in the figure is the graph of $y = f(x)$, $f(x)$ is continuous at all points except at $x_0 = 1$ and

* Jordan, in his *Cours d'analyse*, Paris, 1893, perhaps the most complete treatise on the Calculus ever written, says: "Arithmetic and Algebra employ four fundamental operations, addition, subtraction, multiplication, and division. A fifth can be conceived of, consisting in replacing a variable quantity by its limit. It is the introduction of this new operation that characterizes the Calculus."

$x_0 = -3$. Starting from $(1, -2)$, $f(x)$ is continuous for increasing values of x; starting from $(1, 2)$, $f(x)$ is continuous for decreasing values of x. Starting from $(1, 2)$, limit $[f(1+\Delta x) - f(1)] = 4$ when limit $\Delta x = 0$. Starting from $(1, -2)$,

limit $[f(1-\Delta x) - f(1)] = -4$

when limit $\Delta x = 0$. Starting from $(-3, \infty)$,

limit $[f(-3+\Delta x) - f(-3)]$
$\qquad = -\infty$

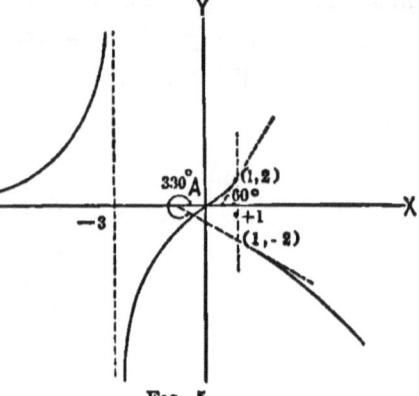

Fig. 5.

when limit $\Delta x = 0$. Starting from $(-3, -\infty)$, limit $[f(-3-\Delta x) - f(-3)] = +\infty$ when limit $\Delta x = 0$.

Starting from $(1, 2)$,

$$\lim \frac{f(1-\Delta x) - f(1)}{-\Delta x} = \tan 60° \quad \text{when limit } \Delta x = 0;$$

$$\lim \frac{f(1+\Delta x) - f(1)}{\Delta x} = \infty \quad \text{when limit } \Delta x = 0.$$

Starting from $(1, -2)$,

$$\lim \frac{f(1+\Delta x) - f(1)}{\Delta x} = \tan 330° \text{ when limit } \Delta x = 0;$$

$$\lim \frac{f(1-\Delta x) - f(1)}{-\Delta x} = \infty \quad \text{when limit } \Delta x = 0.$$

Hence it is evident that, at points of discontinuity of $f(x)$, the limit of the ratio $\dfrac{f(x_0+\Delta x) - f(x_0)}{\Delta x}$ when limit $\Delta x = 0$ is not independent of the algebraic sign of Δx. At points of continuity, the limit of this ratio generally is independent of the sign of Δx.

EXAMPLE. — Show that $\log x$ is a continuous function of x.

Here $\operatorname{limit} [\log (x \pm \Delta x) - \log x] = \operatorname{limit} \log\left(1 \pm \dfrac{\Delta x}{x}\right) = 0$, when $\operatorname{limit} \Delta x = 0$, except for $x = 0$.

PROBLEMS

Show that the following are continuous functions of x:

1. $3x^2 - 5x + 2$. 2. $\sin x$. 3. e^x.

Find the points of discontinuity of

4. $\dfrac{3x+2}{5x-7}$. 5. $\tan x$.

CHAPTER III

DIFFERENTIATION AND INTEGRATION OF ALGEBRAIC FUNCTIONS

Art. 12. — Differentiation

The function of x which is the limit of the ratio

$$\frac{f(x + \Delta x) - f(x)}{\Delta x}$$

when limit $\Delta x = 0$, is called the first derivative of $f(x)$ with respect to x, and is denoted by the symbol $\frac{d}{dx} f(x)$, or $f'(x)$. If $f(x)$ is denoted by y, the first derivative is denoted by $\frac{dy}{dx}$.*
The operation of forming the first derivative, denoted by the symbol $\frac{d}{dx}$, is called differentiation. General rules for the differentiation of algebraic functions are to be established.

I. Let u represent any continuous function of x. Represent by Δu the change in the value of u corresponding to a change of Δx in the value of x. By definition, $\frac{du}{dx} = $ limit $\frac{\Delta u}{\Delta x}$ when limit $\Delta x = 0$.

---

* The notation $\frac{dy}{dx}$ was invented by Leibnitz (1646–1716). Newton (1642–1727) denotes $\frac{ds}{dt}$ by $\dot{s}$, a notation still used in mechanics. Lagrange (1736–1813) denotes the first derivative of $f(x)$ by $f'(x)$, Cauchy (1789–1857) by $Df(x)$.

II. Let $u = c$, where c is a constant, that is, a change in the value of x does not cause a change in the value of c. Then $\dfrac{du}{dx} = \dfrac{dc}{dx} = \text{limit } \dfrac{\Delta c}{\Delta x} = 0$ when limit $\Delta x = 0$. Conversely, if $\dfrac{du}{dx} \equiv 0$, that is, identically zero, u is independent of x. For $\dfrac{du}{dx} \equiv 0$ means that a change in x causes no change in u. Hence u is independent of x.

III. Let $u = x$. Then $\dfrac{du}{dx} = \dfrac{dx}{dx} = \text{limit } \dfrac{\Delta x}{\Delta x} = 1$ when limit $\Delta x = 0$.

IV. Let $f(x) = c \cdot u$, where c is a constant and u represents a continuous function of x. Forming the first derivative,

$$\frac{d}{dx} f(x) = \text{limit } \frac{c \cdot (u + \Delta u) - c \cdot u}{\Delta x} = \text{limit } c \cdot \frac{\Delta u}{\Delta x} = c \cdot \frac{du}{dx}$$

when limit $\Delta x = 0$. Hence the first derivative of a function multiplied by a constant is the constant times the first derivative of the function.

V. Let $f(x) = u + v - w$, where u, v, w represent continuous functions of x. Denoting by $\Delta u, \Delta v$, and Δw the changes in the values of u, v, and w, corresponding to a change Δx in the value of x,

$$\frac{d}{dx} f(x) = \text{limit } \frac{u + \Delta u + v + \Delta v - w - \Delta w - (u + v - w)}{\Delta x}$$

$$= \text{limit } \frac{\Delta u}{\Delta x} + \text{limit } \frac{\Delta v}{\Delta x} - \text{limit } \frac{\Delta w}{\Delta x} = \frac{du}{dx} + \frac{dv}{dx} - \frac{dw}{dx}$$

when limit $\Delta x = 0$. That is, the first derivative of the algebraic sum of a finite number of functions is the like algebraic sum of the first derivatives of the functions.

If two functions $f(x)$ and $\phi(x)$ have the same first derivative, their difference $f(x) - \phi(x)$ is a constant. By hypothesis,

$$\frac{d}{dx}f(x) - \frac{d}{dx}\phi(x) = \frac{d}{dx}\{f(x) - \phi(x)\} \equiv 0,$$

hence $f(x) - \phi(x) = c$.

VI. Let $f(x) = u \cdot v$, where u and v represent continuous functions of x.

$$\frac{d}{dx}f(x) = \frac{d}{dx}(u \cdot v) = \text{limit}\,\frac{(u + \Delta u) \cdot (v + \Delta v) - u \cdot v}{\Delta x}$$

$$= \text{limit}\,(u + \Delta u) \cdot \frac{\Delta v}{\Delta x} + \text{limit}\,v \cdot \frac{\Delta u}{\Delta x} = u \cdot \frac{dv}{dx} + v \cdot \frac{du}{dx}$$

when limit $\Delta x = 0$. Hence the first derivative of the product of two functions is the first function times the first derivative of the second function plus the second function times the first derivative of the first function.

In like manner it is proved that

$$\frac{d}{dx}(u \cdot v \cdot w) = u \cdot v \cdot \frac{dw}{dx} + u \cdot w \cdot \frac{dv}{dx} + v \cdot w \cdot \frac{du}{dx}.$$

VII. Let $f(x) = \dfrac{u}{v}$, where u and v are continuous functions of x.

$$\frac{d}{dx}f(x) = \frac{d}{dx}\left(\frac{u}{v}\right) = \text{limit}\,\frac{\dfrac{u + \Delta u}{v + \Delta v} - \dfrac{u}{v}}{\Delta x}$$

$$= \text{limit}\,\frac{v \cdot \dfrac{\Delta u}{\Delta x} - u \cdot \dfrac{\Delta v}{\Delta x}}{v^2 + v \cdot \Delta v} = \frac{v \cdot \dfrac{du}{dx} - u \cdot \dfrac{dv}{dx}}{v^2}$$

when limit $\Delta x = 0$. That is, the first derivative of the quotient of two functions is the divisor times the first derivative of the dividend minus the dividend times the first derivative of the divisor, divided by the square of the divisor.

If $f(x) = \dfrac{c}{v}$, where c is a constant, $\dfrac{d}{dx}\left(\dfrac{c}{v}\right) = \dfrac{-c \cdot \dfrac{dv}{dx}}{v^2}$. Hence the first derivative of a fraction whose numerator is constant and whose denominator is a function of x, is the numerator with its sign changed times the first derivative of the denominator divided by the square of the denominator.

VIII. Let $f(x) = u^n$, where u is a continuous function of x, and n is any finite, positive integer.

$$\frac{d}{dx} f(x) = \frac{d}{dx} u^n = \operatorname{limit} \frac{(u + \Delta u)^n - u^n}{\Delta x}$$

$$= \operatorname{limit} \left\{ n \cdot u^{n-1} + \frac{n(n-1)}{\underline{2}} \cdot u^{n-2} \cdot \Delta u + \cdots \right.$$

$$\left. + n \cdot u \cdot (\Delta u)^{n-2} + (\Delta u)^{n-1} \right\} \frac{\Delta u}{\Delta x}$$

$$= n \cdot u^{n-1} \cdot \frac{du}{dx},$$

when limit $\Delta x = 0$. Hence the first derivative of a function affected by a finite, positive, integral exponent is the product of the exponent, the function with its exponent diminished by unity, and the first derivative of the function.

If $f(x) = v = u^{\frac{r}{s}}$, where r and s are finite, positive integers, $v^s = u^r$. Forming the first derivatives of both sides of this equation, $s \cdot v^{s-1} \cdot \dfrac{dv}{dx} = r \cdot u^{r-1} \cdot \dfrac{du}{dx}$. Solving for $\dfrac{dv}{dx}$, there results $\dfrac{dv}{dx} = \dfrac{r}{s} \cdot \dfrac{u^{r-1}}{v^{s-1}} \cdot \dfrac{du}{dx} = \dfrac{r}{s} \cdot \dfrac{u^{r-1}}{u^{r - \frac{r}{s}}} \cdot \dfrac{du}{dx} = \dfrac{r}{s} \cdot u^{\frac{r}{s}-1} \cdot \dfrac{du}{dx}$. Hence

$$\frac{d}{dx} \cdot u^{\frac{r}{s}} = \frac{r}{s} \cdot u^{\frac{r}{s}-1} \cdot \frac{du}{dx};$$

that is, the first derivative of a function affected by a finite, positive, fractional exponent is the product of the exponent, the function with its exponent diminished by unity, and the first derivative of the function.

If $f(x) = u^{-n}$, where n is finite and either integral or fractional, $f(x) = \dfrac{1}{u^n}$, and

$$\frac{d}{dx}f(x) = \frac{d}{dx}\left(\frac{1}{u^n}\right) = \frac{-n \cdot u^{n-1} \cdot \dfrac{du}{dx}}{u^{2n}} = -n \cdot u^{-n-1} \cdot \frac{du}{dx}.$$

Hence the first derivative of a continuous function affected by any finite constant exponent is the product of the exponent, the function with its exponent diminished by unity, and the first derivative of the function.

If $u = x$, $\dfrac{d}{dx}x^n = n \cdot x^{n-1}$.

If y is a continuous function of x, x is also a continuous function of y. From the equation $y = f_1(x)$ an equation of the form $x = f_2(y)$ is obtained. Differentiation gives $\dfrac{dy}{dx}$ and $\dfrac{dx}{dy}$. The relation between these derivatives is to be found. The equation $\dfrac{\Delta y}{\Delta x} \cdot \dfrac{\Delta x}{\Delta y} = 1$ is true for all values of Δx, hence

$$\text{limit } \frac{\Delta y}{\Delta x} \cdot \frac{\Delta x}{\Delta y} = \text{limit}\frac{\Delta y}{\Delta x} \cdot \text{limit}\frac{\Delta x}{\Delta y} = \frac{dy}{dx} \cdot \frac{dx}{dy} = 1,$$

when limit $\Delta x = 0$ and limit $\Delta y = 0$. There results $\dfrac{dy}{dx} = \dfrac{1}{\dfrac{dx}{dy}}$;

that is, the first derivative of y with respect to x is the reciprocal of the first derivative of x with respect to y.

If y is a continuous function of z, $y = f_1(z)$, and z is a continuous function of x, $z = f_2(x)$, y is also a continuous function of x. The derivative $\dfrac{dy}{dx}$ is to be calculated.

The equation $\frac{\Delta y}{\Delta x} = \frac{\Delta y}{\Delta z} \cdot \frac{\Delta z}{\Delta x}$ is true for all values of Δx. When limit $\Delta x = 0$, limit $\frac{\Delta y}{\Delta x} =$ limit $\frac{\Delta y}{\Delta z} \cdot$ limit $\frac{\Delta z}{\Delta x}$; that is, $\frac{dy}{dx} = \frac{dy}{dz} \cdot \frac{dz}{dx}$.

The rules of this article are sufficient for the differentiation of all algebraic functions.

EXAMPLE I. — Form the first derivative of
$$5x^3 - 7x^2 + 12x - 15.$$

$$\frac{d}{dx}(5x^3 - 7x^2 + 12x - 15) = 5\frac{d}{dx}x^3 - 7\frac{d}{dx}x^2 + 12\frac{d}{dx}x - \frac{d}{dx}15$$
$$= 15x^2 - 14x + 12.$$

EXAMPLE II. — Form the first derivative of $(2 - 5x^2)^{\frac{3}{2}}$. This expression has the form u^n whose derivative is
$$\frac{d}{dx}u^n = n \cdot u^{n-1} \cdot \frac{du}{dx}.$$

In the problem $u = 2 - 5x^2$, $n = \frac{3}{2}$. Hence
$$\frac{d}{dx}(2-5x^2)^{\frac{3}{2}} = \frac{3}{2}(2-5x^2)^{\frac{1}{2}} \cdot \frac{d}{dx}(2-5x^2) = -15x(2-5x^2)^{\frac{1}{2}}.$$

EXAMPLE III. — Form the first derivative $\frac{dy}{dx}$ of the implicit function $x^2 + y^2 = 9$.

Forming the first derivative of both sides of this equation, $2x + 2y\frac{dy}{dx} = 0$, whence $\frac{dy}{dx} = -\frac{x}{y}$ and $\frac{dx}{dy} = -\frac{y}{x}$.

EXAMPLE IV. — If $\frac{dy}{dz} = \frac{z^3}{(z^2+1)^{\frac{3}{2}}}$ and $x^2 = z^2 + 1$, form $\frac{dy}{dx}$.

From $x^2 = z^2 + 1$, $\frac{dz}{dx} = \frac{x}{z}$. Hence

$$\frac{dy}{dx} = \frac{dy}{dz} \cdot \frac{dz}{dx} = \frac{z^3}{(z^2+1)^{\frac{2}{3}}} \cdot \frac{x}{z} = \frac{z^2 \cdot x}{(z^2+1)^{\frac{2}{3}}} = \frac{(x^2-1) \cdot x}{x^2} = x - \frac{1}{x}.$$

EXAMPLE V. — Form the first derivative of $y = \dfrac{1-x}{(1+x^2)^{\frac{1}{2}}}$.
Applying the rule for differentiating the quotient of two functions,

$$\frac{dy}{dx} = \frac{(1+x^2)^{\frac{1}{2}} \cdot \dfrac{d}{dx}(1-x) - (1-x) \cdot \dfrac{d}{dx}(1+x^2)^{\frac{1}{2}}}{1+x^2}$$

$$= \frac{-(1+x^2)^{\frac{1}{2}} - x(1-x)(1+x^2)^{-\frac{1}{2}}}{1+x^2}$$

$$= \frac{-1-x^2-x+x^2}{(1+x^2)^{\frac{3}{2}}} = -\frac{1+x}{(1+x^2)^{\frac{3}{2}}}.$$

PROBLEMS

Form the first derivative of,

1. $3x+5$.
2. $2x^2+7x+3$.
3. $3x^5-8$.
4. $\dfrac{1}{1+x}$.
5. $\dfrac{1+x}{1-x}$.
6. $\dfrac{1}{1-x^2}$.
7. $\dfrac{1+x^2}{1-x^2}$.
8. $(1-x)^2$.
9. $(1-x^2)^2$.
10. $(1-x)^{-2}$.
11. $(1-x^2)^{\frac{1}{2}}$.
12. $(3+5x)^{-\frac{1}{2}}$.
13. $\left(\dfrac{x}{1+x}\right)^{\frac{1}{2}}$.
14. $(5x-7x^2)^{\frac{2}{3}}$.
15. $(1-x+x^2)^{\frac{5}{2}}$.
16. $(a+bx^n)^m$.
17. $(a+bx^n)^{\frac{m}{n}}$.

18. Form the first derivatives of $\dfrac{1-x^2}{1+x^2}$ and $\dfrac{2}{1+x^2}$, and find the difference of the functions.

In the following equations y is an implicit function of x. Form $\dfrac{dy}{dx}$.

19. $\dfrac{x^2}{a^2} + \dfrac{y^2}{b^2} = 1.$ 20. $\dfrac{x^2}{a^2} - \dfrac{y^2}{b^2} = 1.$ 21. $y^2 - 2px = 0.$

22. $x^2 y + y - x = 0.$ 23. $x^3 - 3xy + y^3 = 0.$

Determine the rate of change of function and variable of

24. $f(x) = 2x - x^2.$ 26. $f(x) = (1-x^2)^{\frac{1}{2}}$ at $x = \frac{1}{2}.$

25. $f(x) = \dfrac{x}{1-x^2}.$ 27. $f(x) = (4-x^2)^{\frac{1}{2}}$ at $x = 2.$

28. Find at what point of the circle $x^2 + y^2 = 9$ the ordinate increases twice as fast as the abscissa.

29. Discuss the rate of change of ordinate and abscissa of $\dfrac{x^2}{a^2} + \dfrac{y^2}{b^2} = 1$ for different points of the ellipse.

The rate of change of ordinate and abscissa is measured by $\dfrac{dy}{dx} = -\dfrac{b^2 x}{a^2 y}.$ When x and y have like signs the ordinate is a decreasing function of x; when x and y have unlike signs the ordinate is an increasing function of x; when $x = 0$, the ordinate is not changing value; when $y = 0$, the ordinate changes infinitely more rapidly than the abscissa. These results agree with the results obtained by examining the ellipse.

30. Determine the rate of change of area and side of an equilateral triangle. Area $= \dfrac{\sqrt{5}}{4} x^2$, x being a side.

31. Find the rate of change of area and radius of a circle when the radius is 10.

32. A man walks on level ground towards a tower 80 feet high. When 60 feet from the foot of the tower find the relative rate of approaching the top and foot of the tower.

Calling the man's distance from the top of the tower y, his distance from the foot of the tower x, $y^2 = x^2 + 6400$.

Find the slopes of the tangents to the following curves at the point (x, y) of the curve:

33. $y = 4x^2$.   34. $y = 6x - x^2$.   35. $\dfrac{x^2}{a^2} - \dfrac{y^2}{b^2} = 1$.

36. Find the slope of the tangent to $y = 8x - x^2$ at $x = 1$.

37. Find where the point generating the circle $x^2 + y^2 = 4$ tends to move parallel to the X-axis. Where parallel to the Y-axis.

Determine the velocity at time t supposing the relation between s the distance and t the time to be expressed by the following equations, where a and u are constants:

38. $s = \frac{1}{2}at^2$.   39. $s = ut + \frac{1}{2}at^2$.   40. $s = ut - \frac{1}{2}at^2$.

41. If $\dfrac{dy}{dz} = z^2(1+z)^{\frac{1}{2}}$ and $1 + z = x^2$, find $\dfrac{dy}{dx}$.

42. If $\dfrac{dy}{dz} = \dfrac{z}{\sqrt{1+z}}$ and $1 + z = x^2$, find $\dfrac{dy}{dx}$.

Art. 13. — Integration

The difference between two functions which have the same first derivative is constant. Hence, if the first derivative of a function is known, the function itself is known, except for an additive arbitrary constant. The process of obtaining a function $f(x)$ from its first derivative $\dfrac{d}{dx} f(x)$ is called integration, and is denoted by the symbol $\int$. The operations denoted by the symbols $\dfrac{d}{dx}$ and $\int$ neutralize each other; that is,

$$\int \dfrac{d}{dx} f(x) = f(x) + C \quad \text{and} \quad \dfrac{d}{dx} \int f(x) = f(x),$$

where C is an arbitrary constant. Hence the rules of integration are at once inferred from the rules of differentiation. The result of integration is called an integral, and the integral is correct if the derivative of the integral is the expression which was integrated.

The rules of algebraic integration, obtained from the rules of algebraic differentiation of Art. 12, are expressed by the formulas:

I. $\int \dfrac{du}{dx} = u + C.$
III. $\int 1 = x + C.$

II. $\int 0 = C.$
IV. $\int a \cdot \dfrac{du}{dx} = a \cdot u + C.$

V. $\int \left(\dfrac{du}{dx} + \dfrac{dv}{dx} - \dfrac{dw}{dx} \right) = u + v - w + C.$

VI. $\int \left(u \cdot \dfrac{dv}{dx} + v \cdot \dfrac{du}{dx} \right) = u \cdot v + C.$

VII. $\int \dfrac{v \cdot \dfrac{du}{dx} - u \cdot \dfrac{dv}{dx}}{v^2} = \dfrac{u}{v} + C.$

VIII. $\int u^n \cdot \dfrac{du}{dx} = \dfrac{u^{n+1}}{n+1} + C.$ When $u = x$, this formula becomes $\int x^n = \dfrac{x^{n+1}}{n+1} + C.$ When $n = -1$, this result is absurd.

The rule of formula VIII. may be stated, the integral of any function affected by an exponent other than -1, multiplied by the first derivative of the function, is the function with its exponent increased by unity divided by the increased exponent, plus an arbitrary constant.

DIFFERENTIATION AND INTEGRATION 31

Since $\int a \cdot \frac{du}{dx} = a \cdot u + C$ and $a \int \frac{du}{dx} = a \cdot u + C$, a constant factor may be shifted from one side of the sign of integration to the other, without affecting the result.

By formula V. the integral of the algebraic sum of a finite number of functions is the like algebraic sum of the integrals of the functions.

EXAMPLE I. — Find the function of x whose first derivative is $4x^2 - 5x$. Denoting the function by $f(x)$, $\frac{d}{dx} f(x) = 4x^2 - 5x$ and $f(x) = \int (4x^2 - 5x) = 4 \int x^2 - 5 \int x = \frac{4}{3}x^3 - \frac{5}{2}x^2 + C$, where the arbitrary constant C is called the constant of integration. This integral is called the indefinite integral. If $f(x)$ is required to have a given value for a given value of x, for example, if $f(x) = 10$ when $x = 1$, the value of C is found from the equation $f(x) = \frac{4}{3}x^3 - \frac{3}{2}x^2 + C$ to be $11\frac{1}{6}$. Hence under the given conditions $f(x) = \frac{4}{3}x^3 - \frac{5}{2}x^2 + 11\frac{1}{6}$. This result is called the corrected integral.

EXAMPLE II. — Find the function of x whose first derivative is $x(1+x^2)^{\frac{3}{2}}$. Denoting the integral by $f(x)$,

$$f(x) = \int x(1+x^2)^{\frac{3}{2}}.$$

This integration can be performed by means of the formula $\int u^n \cdot \frac{du}{dx} = \frac{u^{n+1}}{n+1} + C$ if $x(1+x^2)^{\frac{3}{2}}$ can be separated into factors of the form u^n and $\frac{du}{dx}$. Placing $u = 1 + x^2$, $\frac{du}{dx} = 2x$. Hence

$$f(x) = \tfrac{1}{2} \int (1+x^2)^{\frac{3}{2}} \frac{d}{dx}(1+x^2) = \tfrac{1}{5}(1+x^2)^{\frac{5}{2}} + C.$$

EXAMPLE III. — If $\dfrac{dy}{dx} = 10(x-2)(x^2-4x)^{-\frac{1}{2}}$, find y.

Placing $u = x^2 - 4x$, $\dfrac{du}{dx} = 2x - 4$,

and $\quad y = 10 \int (x-2)(x^2-4x)^{-\frac{1}{2}}$

$\quad\quad = 5 \int (x^2-4x)^{-\frac{1}{2}} \dfrac{d}{dx}(x^2-4x) = 10(x^2-4x)^{\frac{1}{2}} + C.$

EXAMPLE IV. — Determine the equation of a curve such that the slope of the tangent to the curve at any point (x, y) is the negative ratio of the abscissa to the ordinate.

The condition of the problem is expressed by the equation $\dfrac{dy}{dx} = -\dfrac{x}{y}$, whence $y\dfrac{dy}{dx} = -x$. Integrating, $\dfrac{1}{2}y^2 = -\dfrac{1}{2}x^2 + C$ or $x^2 + y^2 = 2C$. This equation represents any circle with center at the origin of coordinates. If the circle is required to contain the point $(3, 4)$, the value of $2C$ must be 25, and the problem has the determinate solution $x^2 + y^2 = 25$.

PROBLEMS

Find the $f(x)$ whose first derivative is,

1. $2 + x$, knowing that $f(x) = 7$ when $x = 2$.
2. $3 - 5x^2$, knowing that $f(x) = 20$ when $x = 5$.
3. $1 + x + x^2 + x^3$, knowing that $f(x) = 12$ when $x = 1$.
4. $x^{\frac{1}{2}}$. 5. $3x^{-\frac{1}{2}}$. 6. $x^{\frac{3}{2}} + 5$. 7. $2x^{\frac{5}{3}} - x^{\frac{2}{3}}$.
8. $(1-x)(2+x^2)$. Multiply out and integrate term by term.
9. $(3x-5)(2x-3x^2)$.
10. $(1+x^2)(3x^2+2)$.
11. $(4x-5)(2x^2-5x)^{\frac{3}{4}}$.
12. $(2x+3)(x^2+3x)^{\frac{1}{3}}$.
13. $(3x^2-10x)(x^3-5x^2)^{\frac{3}{4}}$.
14. $(2x+5)(x^2+5x-7)^{\frac{2}{3}}$.

15. $(1+9x)^{\frac{1}{2}}$.

16. $(a+bx)^{\frac{3}{4}}$.

17. $15(3x^2-8x)(3x^3-12x^2)^{\frac{1}{5}}$.

18. $(x^2+3x-5)(2x^3+9x^2-30x)^{\frac{2}{3}}$.

Find the equation $y = f(x)$ of a curve such that the slope of the tangent at any point (x, y) is,

19. $3x - 7$.

20. $2x + 5$, the curve passing through $(5, 0)$.

21. $x^2 + 5x$, the curve passing through $(0, 0)$.

22. $\dfrac{b^2}{a^2} \cdot \dfrac{x}{y}$, the curve passing through $(a, 0)$.

23. For the cable of a suspension bridge with load uniformly distributed over the horizontal, $\dfrac{dy}{dx} = \dfrac{w}{t_0} \cdot x$, where w is the uniform load per horizontal linear unit and t_0 is the tension at the lowest point. Find equation of curve assumed by cable.

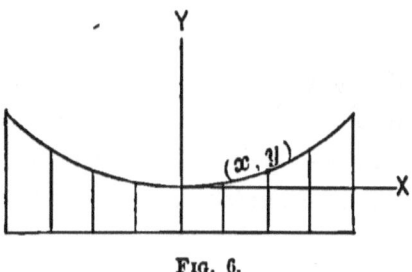

Fig. 6.

24. Find the function of x whose rate of change is $2x - 5$.

Find the relation between s and t, knowing that $s = 0$ when $t = 0$, and that the velocity is,

25. $u + at$.

26. $u - at$.

Find the relation between x and y, knowing that,

27. $\dfrac{dy}{dx} = \dfrac{x^2}{y}$.

28. $\dfrac{dy}{dx} = \dfrac{1+x}{1-y}$.

29. $\dfrac{dy}{dx} = \dfrac{x}{1+y^2}$.

30. $\dfrac{dy}{dx} = \dfrac{1-3x+5x^2}{y-4y^3}$.

31. $\dfrac{dy}{dx} = \dfrac{1}{x^2 y^3}$.

32. $x^3 \dfrac{dy}{dx} = \dfrac{5}{y}$.

Find the values of the following integrals:

33. $\displaystyle\int \frac{x}{\sqrt{1-x^2}}.$ 34. $\displaystyle\int \frac{x}{(1-x^2)^{\frac{3}{2}}}.$ 35. $\displaystyle\int (1-x^2)^3.$

36. $\displaystyle\int \left(\frac{1+x^2}{x}\right)^2.$ 37. $\displaystyle\int \frac{x}{(2ax-x^2)^{\frac{3}{2}}}.$

Multiplying numerator and denominator by $(x^{-2})^{\frac{3}{2}} = x^{-3}$, the problem becomes

$$\int \frac{x^{-2}}{(2ax^{-1}-1)^{\frac{3}{2}}} = -\frac{1}{2a}\int (2ax^{-1}-1)^{-\frac{3}{2}}\frac{d}{dx}(2ax^{-1}-1)$$

$$= \frac{1}{a}(2ax^{-1}-1)^{-\frac{1}{2}} + C = \frac{x}{a\sqrt{2ax-x^2}} + C.$$

38. $\displaystyle\int \frac{\sqrt{2ax-x^2}}{x^3}.$ 39. $\displaystyle\int \frac{1}{(a^2+x^2)^{\frac{3}{2}}}.$ 40. $\displaystyle\int \frac{x}{\sqrt{1+x}}.$

Calling the integral y, the problem is, given $\dfrac{dy}{dx}$, to find y. By the substitution $1 + x = z^2$,

$$\frac{dy}{dz} = \frac{dy}{dx}\cdot\frac{dx}{dz} = \frac{x}{\sqrt{1+x}}\cdot 2z = \frac{2z(z^2-1)}{z} = 2z^2 - 2.$$

By integration $y = \dfrac{2}{3}z^3 - 2z + C,$

hence $\qquad y = \dfrac{2}{3}(1+x)^{\frac{3}{2}} - 2(1+x)^{\frac{1}{2}} + C.$

41. $\displaystyle\int x^2(1+x)^{\frac{1}{2}}.$ 42. $\displaystyle\int \frac{x^3}{(x^2+1)^{\frac{3}{2}}}.$

Art. 14. — Definite Integrals

If $\dfrac{d}{dx}F(x) = f(x)$, $\displaystyle\int f(x) = F(x) + C$. If $F(x) + C$ is evaluated for $x = b$ and $x = a$ and the second result subtracted from the first, the final result $F(b) - F(a)$ is called the definite integral of $f(x)$ between the limits $x = a$ and $x = b$. This operation is denoted by writing

$$\int_a^b f(x) = \Big[F(x) + C\Big]_{x=a}^{x=b} = F(b) - F(a),$$

where a is called the lower limit and b the upper limit of the definite integral.

From this definition it follows that

$$\int_a^b f(x) = -\int_b^a f(x),$$

and also that, if $b - a = (b - c) + (c - a)$,

$$\int_a^b f(x) = \int_c^b f(x) + \int_a^c f(x).$$

EXAMPLE. — Find the value of the definite integral

$$\int_3^5 (3x^2 - 5x + 7).$$

$$\int_3^5 (3x^2 - 5x + 7) = \Big[x^3 - \tfrac{5}{2}x^2 + 7x + C\Big]_{x=3}^{x=5} = 63.$$

PROBLEMS

Calculate the definite integrals,

1. $\int_0^2 (3x^2 - 5).$

2. $\int_2^5 (2x^2 + 4x - 6).$

3. $\int_{-1}^{+1} x(4 - x^2).$

4. $\int_2^4 (1 + x)(1 - x^2).$

5. $\int_0^3 (1 - x^2).$

6. $\int_0^2 x(4 - x^2)^{\frac{1}{2}}.$

7. $\int_0^1 x(1 - x^2)^{\frac{1}{2}}.$

8. $\int_0^a x(a^2 - x^2)^{\frac{1}{2}}.$

9. $\int_0^a x^{\frac{1}{2}} dx.$

10. $\int_0^8 3 x^{\frac{2}{3}} dx.$

11. $\int_0^3 x^2 \sqrt{1+x}$. Write $\dfrac{du}{dx} = x^2 \sqrt{1+x}$ and determine $\dfrac{du}{dz}$ when $1+x = z^2$. There results $\dfrac{du}{dz} = \dfrac{du}{dx} \cdot \dfrac{dx}{dz} = 2\, z^2(z^2-1)^2$. Since $z = 1$ when $x = 0$, and $z = 2$ when $x = 3$,

$$u = \int_0^3 x^2 \sqrt{1+x} = \int_1^2 2\, z^2(z^2-1)^2 = 16.1.$$

12. If s is the distance in feet and t the time in seconds of a body's motion, find the distance the body moves from the end of the third to the end of the fifth second, knowing that $\dfrac{ds}{dt} = 32.16\, t$.

Art. 15.—Evaluation of the Limit of the Sum

The value of $S = \mathrm{limit} \sum\limits_{x=a}^{x=b} f(x) \cdot \Delta x$ when limit $\Delta x = 0$, where $\sum\limits_{x=a}^{x=b} f(x) \cdot \Delta x$ stands for

$$f(a)\cdot\Delta x + f(a+\Delta x)\cdot\Delta x + \cdots + f(b-2\Delta x)\cdot\Delta x + f(b-\Delta x)\cdot\Delta x,$$

is to be determined. Suppose $F(x)$ to represent a continuous function of x whose first derivative is $f(x)$, that is, $\dfrac{d}{dx} F(x) = f(x)$. This means that $\dfrac{F(x+\Delta x) - F(x)}{\Delta x} = f(x) \pm \epsilon$, where ϵ becomes indefinitely small when Δx becomes indefinitely small. From the last equation, $f(x) \cdot \Delta x = F(x+\Delta x) - F(x) \pm \epsilon \cdot \Delta x$. Whence by substituting for x successively a, $a+\Delta x$, $a+2\Delta x$, $\ldots$, $b-2\Delta x$, $b-\Delta x$,

$$
\begin{aligned}
f(a)\cdot\Delta x &= F(a+\Delta x) - F(a) \pm \epsilon_1 \cdot \Delta x,\\
f(a+\Delta x)\cdot\Delta x &= F(a+2\Delta x) - F(a+\Delta x) \pm \epsilon_2 \cdot \Delta x,\\
f(a+2\Delta x)\cdot\Delta x &= F(a+3\Delta x) - F(a+2\Delta x) \pm \epsilon_3 \cdot \Delta x,\\
&\cdots\cdots\cdots\cdots\cdots\cdots\cdots\\
f(b-2\Delta x)\cdot\Delta x &= F(b-\Delta x) - F(b-2\Delta x) \pm \epsilon_{n-1} \cdot \Delta x,\\
f(b-\Delta x)\cdot\Delta x &= F(b) - F(b-\Delta x) \pm \epsilon_n \cdot \Delta x.
\end{aligned}
$$

By addition, $\sum_{x=a}^{x=b} f(x) \cdot \Delta x = F(b) - F(a) \pm \Sigma \epsilon \cdot \Delta x$. Since $F(x)$ is continuous, each ϵ becomes indefinitely small when Δx becomes indefinitely small. Denoting by E the numerically largest ϵ, $\Sigma \epsilon \cdot \Delta x \lessgtr E \Sigma \Delta x = E(b-a)$, which becomes indefinitely small when E becomes indefinitely small, since $b-a$ is supposed to be finite. Hence

$$S = \text{limit} \sum_{x=a}^{x=b} f(x) \cdot \Delta x = F(b) - F(a),$$

when limit $\Delta x = 0$. But $F(b) - F(a)$ is the value of the definite integral $\int_a^b f(x)$, for by hypothesis $\frac{d}{dx} F(x) = f(x)$. Hence, finally, if $\frac{d}{dx} F(x) = f(x)$, $S = \text{limit} \sum_{x=a}^{x=b} f(x) \Delta x = \int_a^b f(x)$ when limit $\Delta x = 0$.

If the summation limits are a and x, the result obtained becomes $S = \int_a^x f(x) = F(x) - F(a)$. Forming the derivative of S, $\frac{dS}{dx} = \frac{d}{dx} F(x) = f(x)$.

Art. 16. — Infinitesimals and Differentials

A quantity which becomes indefinitely small is called an infinitesimal.

If several infinitesimals occur in the same problem, any one may be chosen as the principal infinitesimal. Denote the principal infinitesimal by α.

Infinitesimals whose ratio to α is finite are said to be of the first order.

An infinitesimal β whose ratio to the nth power of α is finite is said to be of the nth order. If n is positive and larger than unity, β is said to be of a higher order than α.

Denoting the finite ratio of β to α^n by r, $\frac{\beta}{\alpha^n} = \frac{\beta}{\alpha} \cdot \frac{1}{\alpha^{n-1}} = r$, whence $\frac{\beta}{\alpha} = r \cdot \alpha^{n-1}$, an infinitesimal, and $\frac{\alpha}{\beta} = \frac{1}{r \cdot \alpha^{n-1}}$, an indefinitely large quantity. Hence, if β is an infinitesimal of a higher order than α, $\beta = \epsilon \cdot \alpha$, where ϵ is an infinitesimal.

The laws governing the use of infinitesimals are contained in the following two propositions.

I. In the limit of the ratio of infinitesimals any infinitesimal may be replaced by another infinitesimal differing from it by infinitesimals of higher orders.

Let α and β represent any two infinitesimals, and consider the ratio

$$r = \frac{A \cdot \alpha + B \cdot \alpha^2 + C \cdot \alpha^3 + D \cdot \alpha^4 + \cdots}{A_1 \cdot \beta + B_1 \cdot \beta^2 + C_1 \cdot \beta^3 + D_1 \cdot \beta^4 + \cdots}$$

$$= \frac{\alpha}{\beta} \frac{A + B \cdot \alpha + C \cdot \alpha^2 + D \cdot \alpha^3 + \cdots}{A_1 + B_1 \cdot \beta + C_1 \cdot \beta^2 + D_1 \cdot \beta^3 + \cdots},$$

where the coefficients $B, C, D, \cdots$ and $B_1, C_1, D_1, \cdots$ are finite. Let M be the numerically largest of the coefficients $B, C, D, \cdots$, M_1 the numerically largest of the coefficients $B_1, C_1, D_1, \cdots$. Then $B \cdot \alpha + C \cdot \alpha^2 + D \cdot \alpha^3 + \cdots \not> M(\alpha + \alpha^2 + \alpha^3 + \cdots) = M\frac{\alpha}{1-\alpha}$, an infinitesimal, and

$$B_1 \cdot \beta + C_1 \cdot \beta^2 + D_1 \cdot \beta^3 + \cdots \not> M_1(\beta + \beta^2 + \beta^3 + \cdots) = M_1 \frac{\beta}{1-\beta},$$

another infinitesimal. Hence in the limit

$$r = \frac{\alpha}{\beta} \frac{A + B \cdot \alpha + C \cdot \alpha^2 + D \cdot \alpha^3 + \cdots}{A_1 + B_1 \cdot \beta + C_1 \cdot \beta^2 + D_1 \cdot \beta^3 + \cdots} = \frac{A\alpha}{B\beta}.$$

II. In the limit of the sum of infinitesimals, provided this limit is finite, any infinitesimal may be replaced by another differing from it by an infinitesimal of a higher order.

Suppose limit $(\alpha_1 + \alpha_2 + \alpha_3 + \cdots) = c$, a finite quantity, and let $\beta_1 = \alpha_1 + \epsilon_1 \cdot \alpha_1$, $\beta_2 = \alpha_2 + \epsilon_2 \cdot \alpha_2$, $\beta_3 = \alpha_3 + \epsilon_3 \cdot \alpha_3$, $\cdots$, where

$\epsilon_1, \epsilon_2, \epsilon_3, \cdots$ are infinitesimals of which ϵ is the numerically largest. Adding

$$(\beta_1 + \beta_2 + \beta_3 + \cdots)$$
$$= (\alpha_1 + \alpha_2 + \alpha_3 + \cdots) + (\epsilon_1 \cdot \alpha_1 + \epsilon_2 \cdot \alpha_2 + \epsilon_3 \cdot \alpha_3 + \cdots).$$

Now $\epsilon_1 \cdot \alpha_1 + \epsilon_2 \cdot \alpha_2 + \epsilon_3 \cdot \alpha_3 + \cdots \not> \epsilon(\alpha_1 + \alpha_2 + \alpha_2 + \cdots) = \epsilon \cdot c$, an infinitesimal. Hence in the limit

$$\beta_1 + \beta_2 + \beta_3 + \cdots = \alpha_1 + \alpha_2 + \alpha_3 + \cdots.$$

This proposition is true even when c is indefinitely large, provided $\epsilon \cdot c$ is an infinitesimal.

If the limit of the ratio of two infinitesimals is unity, the infinitesimals can differ only by infinitesimals of higher orders. For if limit $\frac{\beta}{\alpha} = 1$, $\frac{\beta}{\alpha} = 1 \pm \epsilon$ and $\beta = \alpha \pm \epsilon \cdot \alpha$, where ϵ is an infinitesimal. Hence the rules governing the use of infinitesimals may be stated, in the limit of the ratio and in the limit of the sum any infinitesimal may be replaced by another infinitesimal, provided the limit of the ratio of the two infinitesimals is unity.

Since infinitesimals of higher orders disappear from the limit of the ratio and from the limit of the sum, the solution of problems involving the limit of the ratio or the limit of the sum may be simplified by dropping infinitesimals of higher orders at the start.

If $y = f(x)$ is a continuous function of x whose first derivative is $f'x$, $\frac{\Delta y}{\Delta x} = f'(x) + \epsilon$, whence $\Delta y = f'x \cdot \Delta x + \epsilon \cdot \Delta x$, where Δy is the difference in the value of the function corresponding to a difference of Δx in the value of the variable, and ϵ becomes an infinitesimal when Δx becomes an infinitesimal. When the difference Δx becomes an infinitesimal it is denoted by dx, which is read differential x. The change in the value of y corresponding to dx is $f'(x) \cdot dx + \epsilon \cdot dx$. Defining dy, read differential y, by the equation $dy = f'(x) \cdot dx$, dy differs from the change in the value of y corresponding to a change of dx in the value of x by $\epsilon \cdot dx$, an infinitesimal of a higher order than dx. Hence in problems involving the limit of the ratio or the limit of the sum the actual change in y may be replaced by $dy = f'(x) \cdot dx$.

The first derivative of a function is therefore the factor by which the differential of the variable must be multiplied to obtain the corresponding differential of the function. This explains why the first derivative of a function is frequently called the differential coefficient of the function.

The operation of finding the differential of a function corresponding to the differential of the variable is called differentiation. It will be observed that differentiation as here defined is performed by the rules established for differentiation as defined in Art. 12.

For example, if $y = x^3 - 7x^2 + 15x + 10$,

$$dy = \frac{d}{dx}(x^3 - 7x^2 + 15x + 10)\, dx = (3x^2 - 14x + 15)\, dx.$$

If $F(x)$ is a continuous function of x whose first derivative is $f(x)$, $\Delta F(x) = F(x + \Delta x) - F(x) = f(x) \cdot \Delta x + \epsilon \cdot \Delta x$, where $\Delta F(x)$ and ϵ become infinitesimals when Δx becomes an infinitesimal. The sum of the elements $\Delta F(x)$ of $F(x)$ from $x = a$ to $x = x$ is $F(x) - F(a) = \sum_{x=a}^{x=x} [f(x) \cdot \Delta x + \epsilon \cdot \Delta x]$, and the sum of the elements Δx of x from $x = a$ to $x = x$ is $x - a$. This is true for all magnitudes of Δx. When the element Δx becomes the infinitesimal element dx,

$$F(x) - F(a) = \sum_{x=a}^{x=x} [f(x) \cdot dx + \epsilon \cdot dx].$$

Supposing $x - a$ to be finite, in the limit

$$F(x) - F(a) = \sum_{x=a}^{x=x} f(x) \cdot dx,$$

since $\epsilon \cdot dx$ is an infinitesimal of a higher order than dx. Calling the operation of finding the limit of the sum integration and denoting it by the symbol $\int$, and indicating that the sum is to extend from $x = a$ to $x = x$ by writing $\int_a^x$, the result obtained may be written $\int_a^x f(x) \cdot dx = F(x) - F(a)$. Now $F(x) - F(a)$ was found in Art. 14 to be the value of the definite integral $\int_a^x f(x)$, provided $\frac{d}{dx} F(x) = f(x)$.

In general, if $y = f(x)$ and $\frac{dy}{dx} = f'(x)$, in the notation of differentials $dy = f'(x) \cdot dx$ and $y = \int f'(x) \cdot dx = f(x) + C$. Differentiation and integration as defined in this article are again inverse operations.

It will be observed that integration as here defined is performed by the rules established for integration in Art. 13. For example, if

$$dy = (3x^2 - 5x + 10) dx, \quad y = x^3 - \tfrac{5}{2}x^2 + 10x + C;$$

the sum of the elements of y from $x = 0$ to $x = x$ is

$$\int_0^x (3x^2 - 5x + 10) dx = \left[x^3 - \tfrac{5}{2}x^2 + 10x + C \right]_0^x$$
$$= x^3 - \tfrac{5}{2}x^2 + 10x;$$

the sum of the elements of y from $x = 0$ to $x = 4$ is

$$\int_0^4 (3x^2 - 5x + 10) dx = \left[x^3 - \tfrac{5}{2}x^2 + 10x + C \right]_0^4 = 64.$$

While the method of differentials is based on the method of limits, the method of differentials has two decided advantages: first, calculations are simplified by dropping differentials of higher order at the start; secondly, the successive steps in the application of the method of differentials, especially in summation problems, are more directly intelligible than is the case when using the method of limits. Hereafter the derivative and differential notations are used interchangeably.

PROBLEMS

Form the differentials of the functions,

1. $y = 3x^2 - 5$.

2. $y = (1 - x^2)^{\frac{1}{2}}$.

3. $y = \dfrac{x^2 - 1}{x^2 + 1}$.

4. $y = \dfrac{1}{(1-x)^2}$.

Find the functions whose differentials are,

5. $dy = (2x^3 - 5x)\,dx$.

6. $dy = (1 + x^2)\,dx$.

7. $dy = (1 + x)^3\,dx$.

Evaluate the definite integrals,

8. $\displaystyle\int_0^5 (4x - 5)\,dx$.

9. $\displaystyle\int_2^4 (2x^3 - 6x)\,dx$.

10. $\displaystyle\int_5^7 (x^2 - 3x + 5)\,dx$.

11. $\displaystyle\int_0^3 x^3(x^2 + 1)^{-\frac{3}{2}}\,dx$. Placing $x^2 + 1 = z^2$, $dx = \dfrac{z}{x}dz$, and

$$x^3(x^2+1)^{-\frac{3}{2}}\,dx = x^3 \cdot z^{-3} \cdot \frac{z}{x} \cdot dz = x^2 \cdot z^{-2} \cdot dz$$

$$= (z^2 - 1) \cdot z^{-2} \cdot dz = dz - \frac{dz}{z^2}.$$

While x takes all values from 0 to 3, z takes all values from 1 to 10. Hence

$$\int_0^3 x^3(x^2+1)^{-\frac{3}{2}}\,dx = \int_1^{10}\left(dz - \frac{dz}{z^2}\right) = \left[z + \frac{1}{z} + C\right]_1^{10} = 8.1.$$

12. $\displaystyle\int_1^4 x^2(1+x)^{\frac{1}{2}}\,dx$.

CHAPTER IV

APPLICATIONS OF ALGEBRAIC DIFFERENTIATION AND INTEGRATION

Art. 17. — Tangents and Normals

The slope of the tangent to the curve whose equation is $y = f(x)$ at any point (x_0, y_0) is the first derivative of $y = f(x)$ evaluated for $x = x_0$, $y = y_0$. This is denoted by writing $\tan \alpha = \dfrac{dy_0}{dx_0}$. The equation of the tangent TT' to $y = f(x)$ at (x_0, y_0) is $y - y_0 = \dfrac{dy_0}{dx_0}(x - x_0)$. The intercept of the tangent on the X-axis, found by placing $y = 0$ in the equation of the tangent and solving for x, is $AT = x_0 - y_0 \dfrac{dx_0}{dy_0}$; the intercept on the Y-axis is $AT' = y_0 - x_0 \dfrac{dy_0}{dx_0}$.

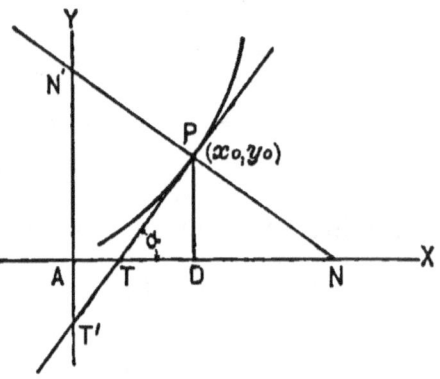

Fig. 7.

The portion of the tangent bounded by the point of tangency and the point of intersection of the tangent with the X-axis is called the length of the tangent. From the figure the length of the tangent is $PT = y_0 \sqrt{1 + \left(\dfrac{dx_0}{dy_0}\right)^2}$. The projection of the tangent PT on the X-axis is called the subtangent. From the figure the subtangent is $DT = y_0 \dfrac{dx_0}{dy_0}$.

The equation of the normal NN' to $y = f(x)$ at (x_0, y_0) is $y - y_0 = -\dfrac{dx_0}{dy_0}(x - x_0)$. The intercept of the normal on the X-axis is $AN = x_0 + y_0 \dfrac{dy_0}{dx_0}$; the intercept of the normal on the Y-axis is $AN' = y_0 + x_0 \dfrac{dx_0}{dy_0}$.

The portion of the normal bounded by the point (x_0, y_0) and the intersection of the normal with the X-axis is called the length of the normal. From the figure the length of the normal is $PN = y_0 \sqrt{1 + \left(\dfrac{dy_0}{dx_0}\right)^2}$. The projection of PN on the X-axis is called the subnormal. From the figure the subnormal is $DN = y_0 \dfrac{dy_0}{dx_0}$.

EXAMPLE. — Find the equations of tangent and normal to $x^2 + 2y^2 - 2xy - x = 0$ at the point $(1, 1)$. Also the length of tangent and subtangent, and of normal and subnormal. Differentiating $x^2 + 2y^2 - 2xy - x = 0$ with respect to x, $2x + 4y\dfrac{dy}{dx} - 2y - 2x\dfrac{dy}{dx} - 1 = 0$, whence $\dfrac{dy}{dx} = \dfrac{1 + 2y - 2x}{4y - 2x}$.
At the point of tangency $x_0 = 1$, $y_0 = 1$, $\dfrac{dy_0}{dx_0} = \dfrac{1}{2}$, $\dfrac{dx_0}{dy_0} = 2$. Hence the equation of the tangent is $y - 1 = \tfrac{1}{2}(x - 1)$, reducing to $y = \tfrac{1}{2}x + \tfrac{1}{2}$; the equation of the normal $y - 1 = -2(x - 1)$, reducing to $y = -2x + 3$; the length of the tangent is $\sqrt{5}$; the length of the subtangent 2; the length of the normal $\tfrac{1}{2}\sqrt{5}$; the length of the subnormal $\tfrac{1}{2}$.

PROBLEMS

1. Find the equations of tangents and normals to $x^2y + y - x = 0$ at $x = +1$ and at $x = -1$.

2. Find the equations of tangent and normal to $xy = 4$ at $x = 2$.

3. Find equation of tangent to $x^{\frac{2}{3}} + y^{\frac{2}{3}} = a^{\frac{2}{3}}$ at (x_0, y_0).

4. Show that the sum of the intercepts of the tangent to $x^{\frac{1}{2}} + y^{\frac{1}{2}} = a^{\frac{1}{2}}$ is the same for all positions of the point of tangency.

5. Find the subtangent of the ellipse $\dfrac{x^2}{a^2} + \dfrac{y^2}{b^2} = 1$.

6. Find the subnormal of the parabola $y^2 = 2px$.

7. Find the length of the normal to $4x^2 + 16y^2 = 100$ at $x = 3$.

8. Find the length of the tangent to $4x^2 + 16y^2 = 100$ at $x = 3$.

9. Find the length of the subnormal to $x^2 + y^2 = 25$ at $(3, 4)$.

10. Determine the curves whose subnormal is constant.

The hypothesis is expressed by the equation $y\dfrac{dy}{dx} = a$, where a is the constant length of the subtangent. From this equation $\dfrac{dx}{dy} = \dfrac{y}{a}$ or $a\,dx = y\,dy$. By integration $ax = \tfrac{1}{2}y^2 + C$ or $y^2 = 2ax - 2C$. This equation represents a system of parabolas.

Art. 18. — Length of a Plane Curve

Denote by s the length from $x = a$ to $x = b$ of the plane curve whose equation is the continuous function $y = f(x)$.

Divide ab into any number of equal parts Δx, and draw ordinates at the points of division of ab. Draw the chords of the arcs Δs into which these ordinates divide the curve AB. At the ends of any one of these arcs, such as mn, draw tangents to the curve AB. Assume as axiomatic that chord $mln <$ arc $mn <$ broken line mkn. From the right triangles klm and kln,

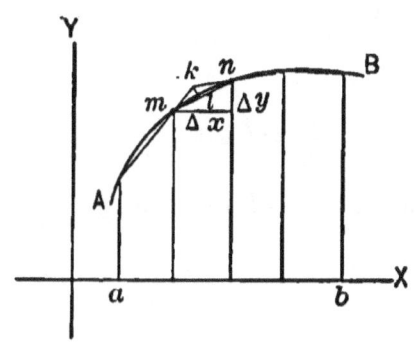

Fig. 8.

$$km = lm \cdot \sec kml,$$
$$kn = ln \cdot \sec knl.$$

These relations are true for all magnitudes of the chord mn. As the length of the chord mn is diminished, the angles kml and knl approach zero and their secants approach unity. Hence $km = lm(1 + \epsilon_1)$ and $kn = ln(1 + \epsilon_2)$, where ϵ_1 and ϵ_2 approach zero when mn approaches zero. Adding,

$$km + kn = (lm + ln) - (\epsilon_1 \cdot lm + \epsilon_2 \cdot ln).$$

When the chord mln becomes an infinitesimal, lm, ln, ϵ_1, and ϵ_2 become infinitesimals. It follows that

$$\text{broken line } mkn - \text{chord } mln = \epsilon_1 \cdot lm + \epsilon_2 \cdot ln,$$

an infinitesimal of a higher order than the chord mln. Hence the difference between the infinitesimal arc mn and its chord mln is an infinitesimal of a higher order than the chord, and in problems involving the limit of the ratio or the limit of the sum, the infinitesimal arc may be replaced by its chord.

It follows at once that the length of the curve is the limit of the length of the inscribed broken line when the number of parts into which ab is divided is indefinitely increased, which is equivalent to saying when limit $\Delta x = 0$. That is,

$$s = \text{limit} \sum_{x=a}^{x=b} (\Delta x^2 + \Delta y^2)^{\frac{1}{2}} = \text{limit} \sum_{x=a}^{x=b} \left(1 + \frac{\Delta y^2}{\Delta x^2}\right)^{\frac{1}{2}} \cdot \Delta x$$

when limit $\Delta x = 0$.

Hence $s = \int_a^b \left(1 + \frac{dy^2}{dx^2}\right)^{\frac{1}{2}} \cdot dx$ and $\dfrac{ds}{dx} = \left(1 + \dfrac{dy^2}{dx^2}\right)^{\frac{1}{2}}$.

Since $\dfrac{dy}{dx} = \tan \alpha$, where α is the angle of inclination to the X-axis of the tangent to the curve $y = f(x)$ at (x, y),

$$\frac{ds}{dx} = (1 + \tan^2 \alpha)^{\frac{1}{2}} = \sec \alpha, \text{ whence } \cos \alpha = \frac{dx}{ds}.$$

Since $\sin \alpha = \tan \alpha \cdot \cos \alpha$, $\sin \alpha = \dfrac{dy}{dx} \cdot \dfrac{dx}{ds} = \dfrac{dy}{ds}$. These results, obtained by the method of limits, may be roughly inferred from the figure. As Δx approaches zero, the element of curve Δs approaches equality with its chord and becomes a part of the tangent. Hence in the limit, when Δx, Δy, and Δs become the infinitesimals dx, dy, and ds,

$$ds^2 = dx^2 + dy^2 = \left(1 + \frac{dy^2}{dx^2}\right) \cdot dx^2, \quad ds = \left(1 + \frac{dy^2}{dx^2}\right)^{\frac{1}{2}} \cdot dx,$$

$$\tan \alpha = \frac{dy}{dx}, \quad \sin \alpha = \frac{dy}{ds}, \quad \cos \alpha = \frac{dx}{ds}.$$

EXAMPLE. — Find the length of the semi-cubic parabola $y^2 = 4x^3$ from $x = 5$ to $x = 10$.

From the equation of the curve, $y = 2x^{\frac{3}{2}}$ and $\dfrac{dy}{dx} = 3x^{\frac{1}{2}}$.

Hence, $s = \int_5^{10} (1 + 9x)^{\frac{1}{2}} \cdot dx = \left[\tfrac{2}{27}(1 + 9x)^{\frac{3}{2}} + C\right]_5^{10} = 31.07$.

The length of the curve from the origin $(0, 0)$ to any point (x, y) is

$$s = \int_0^x (1 + 9x)^{\frac{1}{2}} \cdot dx = \tfrac{2}{27}(1 + 9x)^{\frac{3}{2}} - \tfrac{2}{27}.$$

The length of the curve between any two points $x = a$ and $x = b$ is

$$s = \int_a^b (1 + 9x)^{\frac{1}{2}} \cdot dx = \tfrac{2}{27}(1 + 9b)^{\frac{3}{2}} - \tfrac{2}{27}(1 + 9a)^{\frac{3}{2}}.$$

PROBLEMS

1. Find the length of $y = 2x$ from $x = 5$ to $x = 10$.
2. Find the length of $y = 3x + 5$ from $x = 5$ to $x = 20$.
3. Find the length of $9y^2 = 4x^3$ from $(0, 0)$ to (x, y).
4. Find the length of $9y^2 = 4x^3$ from $x = 0$ to $x = 10$.
5. Find the length of $9y^2 = 4x^3$ from $x = 5$ to $x = 15$.

ART. 19. — AREA OF A PLANE SURFACE

Denote by A the area of the surface bounded by the continuous curve whose equation is $y = f(x)$, the lines $x = a$, $x = b$, and the X-axis. Divide the portion of the X-axis from $x = a$ to $x = b$ into any number of equal parts Δx. Construct rectangles on each Δx and the adjacent ordinates as indicated in the figure. Denote by ΔA the portion of A included between two successive ordinates. Then $\Delta A = y \cdot \Delta x + \theta \cdot \Delta x \cdot \Delta y$, where θ is less than unity. This is true for all magnitudes of Δx. When Δx becomes an infinitesimal, Δy also becomes an infinitesimal. Hence the infinitesimal element of area differs from the area of the infinitesimal rectangle by an infinitesimal of a higher order, and it follows that

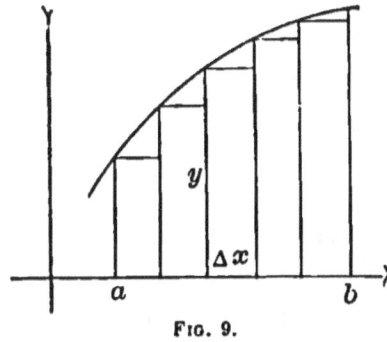

Fig. 9.

$$A = \operatorname{limit} \sum_{x=a}^{x=b} y \cdot \Delta x = \int_a^b y \cdot dx \text{ when limit } \Delta x = 0.$$

APPLICATIONS

EXAMPLE. — Find the area of the surface bounded by the parabola $y^2 = 4x$, the X-axis, and the lines $x = 4$, $x = 9$. For this curve

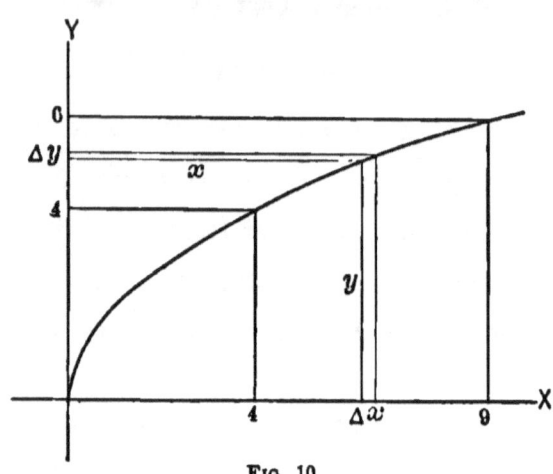

FIG. 10.

$$y = 2x^{\frac{1}{2}} \text{ and } A = 2\int_4^9 x^{\frac{1}{2}} \cdot dx = \tfrac{4}{3}\left[x^{\frac{3}{2}}\right]_4^9 = 28.$$

The area from the vertex to the point (x, y) is

$$A = 2\int_0^x x^{\frac{1}{2}} \cdot dx = \tfrac{4}{3} x^{\frac{3}{2}}.$$

The area from $x = a$ to $x = b$ is

$$A = 2\int_a^b x^{\frac{1}{2}} \cdot dx = \tfrac{4}{3}(b^{\frac{3}{2}} - a^{\frac{3}{2}}).$$

The area of the surface bounded by the parabola $y^2 = 4x$, the Y-axis, and the lines $y = 4$, $y = 6$ is

$$A = \int_4^6 x\,dy = \tfrac{1}{4}\int_4^6 y^2 \cdot dy = \tfrac{1}{12}\left[y^3\right]_4^6 = 12.67.*$$

* The areas of surfaces bounded by curves whose equations are not known may be found mechanically by means of an instrument called the planimeter.

PROBLEMS

1. Find area bounded by $y = 3x$, the lines $x = 0$, $x = 8$ and the X-axis.

2. Find area bounded by $y = 5x$, the lines $x = 1$, $x = 4$ and the X-axis.

3. Find area bounded by $y = mx + n$, the lines $x = a$, $x = b$ and the X-axis.

4. Find area bounded by $y = mx + n$, the lines $y = a$, $y = b$ and the Y-axis.

5. Find area bounded by $y^2 = 9x$, $x = 0$, $x = 4$ and the X-axis.

6. Find area bounded by the parabola $y^2 = 2px$, the ordinate of the point (x, y) of the parabola and the X-axis.

7. Find area bounded by the parabola $y^2 = 2px$, the abscissa of the point (x, y) of the parabola and the Y-axis.

8. Find area bounded by $x^{\frac{1}{2}} + y^{\frac{1}{2}} = a^{\frac{1}{2}}$ and the coordinate axes.

9. Find area bounded by $y^2 = 9x$, $y = x$ and $x = 4$.

10. Find area bounded by $y^2 = 9x$, $y = 2x$ and $y = 6$.

11. Find area bounded by $y^2 = 4x$ and $y = \frac{1}{2}x$.

12. Find area bounded by the parabolas $y^2 = 4x$ and $x^2 = 4y$.

Art. 20. — Area of a Surface of Revolution

Denote by A the area of the surface generated by the revolution of the continuous curve $y = f(x)$ from $x = a$ to $x = b$ about the X-axis. A is the limit, when limit $\Delta x = 0$, of the sum of the areas of the frustums of cones of revolution generated by the revolution of the parts of the broken line ANB. Hence $A = \text{limit} \sum_{x=a}^{x=b} 2\pi \left(\frac{y+y+\Delta y}{2}\right)\left(1+\frac{\Delta y^2}{\Delta x^2}\right)^{\frac{1}{2}} \Delta x$

$$= 2\pi \int_a^b y \left(1+\frac{dy^2}{dx^2}\right)^{\frac{1}{2}} \cdot dx = 2\pi \int_a^b y \cdot ds,$$

when limit $\Delta x = 0$.

EXAMPLE. — Find the area of the surface generated by the revolution of the line $y = 2x$ about the X-axis and bounded by planes perpendicular to the X-axis at $x = 3$ and $x = 8$.

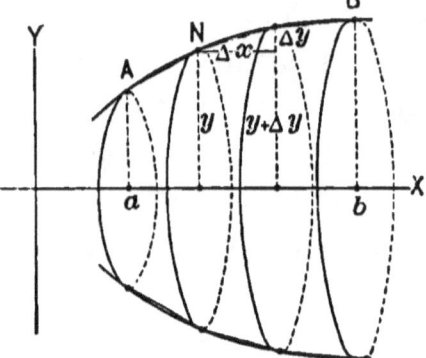

FIG. 11.

Here

$$A = 2\pi \int_3^8 2x(1+4)^{\frac{1}{2}} \cdot dx = 4\sqrt{5}\,\pi \int_3^8 x\,dx = \frac{4\sqrt{5}}{2}\pi \left[x^2\right]_3^8$$
$$= 2\sqrt{5}\,\pi \cdot 55 = 110\sqrt{5}\,\pi = 755.592.$$

PROBLEMS

1. Find the area of the surface generated by the revolution of the line $y = 3x + 5$ about the X-axis and included by the planes perpendicular to the X-axis at $x = 0$, $x = 5$.

2. Find the area of the surface bounded by the revolution of the line $y = c$ about the X-axis and bounded by planes perpendicular to the X-axis at $x = a$, $x = b$.

3. Find the area of the surface of revolution generated by the line $y = mx + n$ revolving about the X-axis and included by planes perpendicular to the X-axis at $x = a$, $x = b$.

4. Find the area of the surface generated by the revolution of $x^{\frac{2}{3}} + y^{\frac{2}{3}} = a^{\frac{2}{3}}$ about the X-axis.

Art. 21. — Volume of a Solid of Revolution

Denote by V the volume of the solid bounded by the surface generated by the revolution of the continuous curve $y = f(x)$ about the X-axis and planes perpendicular to the X-axis at $x = a$, $x = b$.

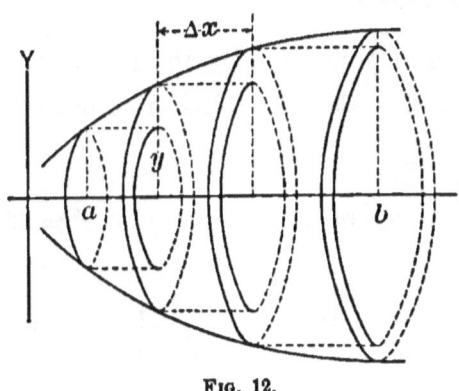

Fig. 12.

Denoting by ΔV the volume of the part of the solid included by planes perpendicular to the X-axis at x and $x + \Delta x$,

$$\pi y^2 \cdot \Delta x < \Delta V < \pi (y + \Delta y)^2 \cdot \Delta x,$$

and
$$\Delta V - \pi y^2 \cdot \Delta x = \theta \pi (2y \cdot \Delta y + \Delta y^2) \Delta x,$$

where θ is less than unity. When Δx becomes an infinitesimal, Δy also becomes an infinitesimal. Hence in the limit when limit $\Delta x = 0$,

$$\Delta V = \pi y^2 \Delta x, \quad \text{and} \quad V = \sum_{x=a}^{x=b} \pi y^2 \Delta x = \pi \int_a^b y^2 \, dx.$$

Example. — Find the volume of the prolate spheroid. The prolate spheroid is generated by the revolution of the ellipse $\dfrac{x^2}{a^2} + \dfrac{y^2}{b^2} = 1$ about the X-axis. The entire spheroid is comprised between $x = +a$ and $x = -a$. Hence,

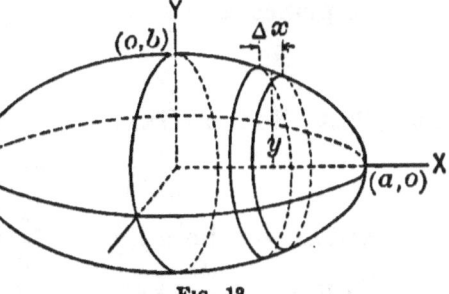

Fig. 13.

APPLICATIONS 53

$$V = \pi \int_{-a}^{+a} \frac{b^2}{a^2}(a^2 - x^2)\, dx = \pi \frac{b^2}{a^2}\left[a^2 x - \tfrac{1}{3}x^3\right]_{-a}^{+a} = \tfrac{4}{3}\pi a b^2.$$

PROBLEMS

1. Find the volume of the solid generated by the revolution of the ellipse $\frac{x^2}{a^2} + \frac{y^2}{b^2} = 1$ about the Y-axis. This is called the oblate spheroid.

2. Find the volume of the part of the sphere $x^2 + y^2 + z^2 = 25$ between $x = 2$ and $x = 4$.

3. Find the volume of the solid bounded by the surface generated by the revolution of the parabola $y^2 = 2px$ about the X-axis and the plane $x = a$.

4. Find the volume of the solid bounded by the surface generated by the revolution of $\frac{x^2}{a^2} - \frac{y^2}{b^2} = 1$ about the X-axis and the planes $x = c$, $x = d$, where c and d are greater than a.

5. Find the volume of the solid bounded by the surface generated by the revolution of $y = 3x + 2$ about the X-axis and planes perpendicular to the X-axis at $x = 2$, $x = 7$.

6. Find the volume of the solid bounded by the surface generated by the revolution of $x^{\frac{1}{2}} + y^{\frac{1}{2}} = a^{\frac{1}{2}}$ about the X-axis and the planes $x = 0$, $x = a$.

ART. 22. — SOLIDS GENERATED BY THE MOTION OF A PLANE FIGURE

EXAMPLE. — The ellipsoid $\frac{x^2}{a^2} + \frac{y^2}{b^2} + \frac{z^2}{c^2} = 1$ may be generated by an ellipse whose center moves on the X-axis, whose plane is perpendicular to the X-axis, and whose axes in any position are the intersections of the plane of the generating ellipse with the fixed ellipses,

$$\frac{x^2}{a^2}+\frac{y^2}{b^2}=1, \ \frac{x^2}{a^2}+\frac{z^2}{c^2}=1.$$

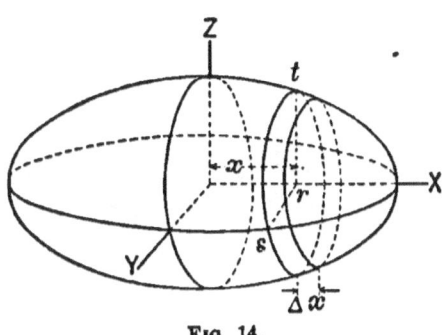

Fig. 14.

The area of the generating ellipse in any position is $\pi \, rs \cdot rt$. Since $(x, rs, 0)$ and $(x, 0, rt)$ are points in the ellipsoid $\frac{x^2}{a^2}+\frac{y^2}{b^2}+\frac{z^2}{c^2}=1$, $rs=\frac{b}{a}(a^2-x^2)^{\frac{1}{2}}$ and $rt=\frac{c}{a}(a^2-x^2)^{\frac{1}{2}}$. Hence, denoting the area of the generating ellipse in terms of x by X,

$$X=\pi\frac{bc}{a^2}(a^2-x^2).$$

Denoting by ΔV the volume of the part of the ellipsoid included by the generating ellipses at x and $x+\Delta x$,

$$X \Delta x > \Delta V > (X - \Delta X)\Delta x,$$

and
$$\Delta V = X \Delta x - \theta \cdot \Delta X \cdot \Delta x,$$

where ΔX represents the change of $X = \pi \frac{bc}{a^2}(a^2-x^2)$ corresponding to a change of Δx in the value of x and θ is less than unity. Since ΔX becomes an infinitesimal when Δx becomes an infinitesimal, in the limit, when limit $\Delta x = 0$, the volume of the ellipsoid is

$$V = \text{limit} \sum_{x=-a}^{x=+a} X \cdot \Delta x = \int_{-a}^{+a} X \cdot dx$$

$$= \pi\frac{bc}{a^2}\int_{-a}^{+a}(a^2-x^2)\,dx = \tfrac{4}{3}\pi\,abc.$$

APPLICATIONS

PROBLEMS

1. Find the volume of the part of the elliptic paraboloid $\frac{y^2}{9} + \frac{z^2}{4} = 2x$ included between the planes $x = 0$ and $x = 5$.

2. Find the volume of the solid bounded by the hyperboloid of one sheet $\frac{x^2}{9} + \frac{y^2}{4} - z^2 = 1$ and the planes $z = 3$, $z = 5$.

3. Two equal semi-circles $x^2 = 2rz - z^2$, $y^2 = 2rz - z^2$ lie in the perpendicular planes XZ, YZ. The solid generated by the square whose center moves on the Z-axis, whose plane is perpendicular to the Z-axis, and whose dimensions in any position are the chords of intersection of the plane of the square with the fixed semi-circles, is called a semi-circular groin. Find the volume of this groin.

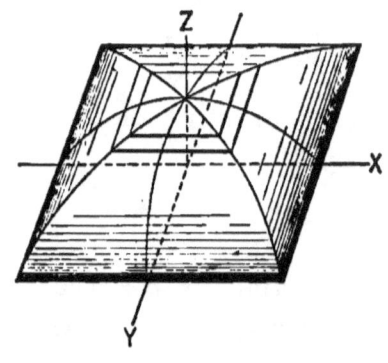

Fig. 15.

Notice that the groin might be defined as the solid which the two cylinders $x^2 = 2rz - z^2$ and $y^2 = 2rz - z^2$ have in common.

4. Two semi-circles $x^2 = 2r_1z - z^2$, $y^2 = 2r_2z - z^2$ of unequal radii lie in the perpendicular planes XZ, YZ. Find the volume of the groin generated by the rectangle whose plane is perpendicular to the Z-axis and whose dimensions are the chords of intersection of the plane of the rectangle with the given semi-circles. Take depth of groin d.

5. The two parabolas $x^2 = 2p_1z$, $y^2 = 2p_2z$ lie in the perpendicular planes XZ, YZ. Find the volume of the groin

generated by the rectangle whose plane is perpendicular to the Z-axis and whose dimensions are the chords of intersection of the plane of the rectangle with the given parabolas. Take depth of groin d.

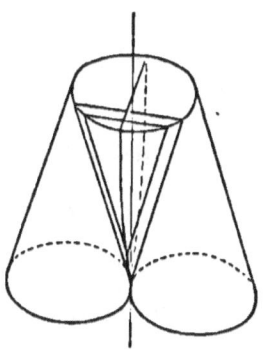

Fig. 16.

6. Two circular cylinders have equal bases and equal altitudes. Their lower bases are tangent to each other, their upper bases coincident. Find the volume of the solid common to the two cylinders.

CHAPTER V

SUCCESSIVE ALGEBRAIC DIFFERENTIATION AND INTEGRATION

Art. 23. — The Second Derivative

The first derivative of $f(x)$, denoted by $\frac{d}{dx}f(x)$ or $f'(x)$, is in general a function of x. The first derivative of the first derivative, $\frac{d}{dx}\left[\frac{d}{dx}f(x)\right]$, is denoted by $\frac{d^2}{dx^2}f(x)$ or $f''(x)$, and is called the second derivative of $f(x)$. If $f(x)$ is denoted by y, the first and second derivatives are written $\frac{dy}{dx}$ and $\frac{d^2y}{dx^2}$.

For example,

if $f(x) = x^3 - 7x + 7$, $f'(x) = 3x^2 - 7$, $f''(x) = 6x$.

Geometrically $\frac{dy}{dx}$ measures the slope of the tangent to the curve whose equation is $y = f(x)$ at (x, y). Hence $\frac{d}{dx}\left(\frac{dy}{dx}\right) = \frac{d^2y}{dx^2}$ measures the rate of change of the slope of the tangent. When $\frac{d^2y}{dx^2}$ is positive, $\frac{dy}{dx}$ or the slope of the tangent increases as x increases. From the figure this is seen to be the case for the part cde of the curve, that is, when the curve is concave

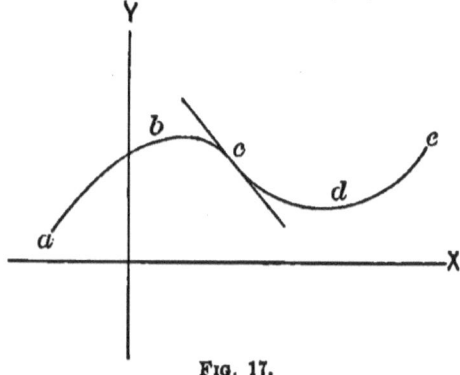

Fig. 17.

upward. When $\frac{d^2y}{dx^2}$ is negative, the slope of the tangent decreases as x increases. This is the case for the part abc of the curve; that is, when the curve is convex upward. When $\frac{d^2y}{dx^2} = 0$, the tangent is stationary. A point of the curve where the tangent is stationary is called a point of inflection, since at such a point the curve changes its direction of curvature.

If the relation between the distance passed over and the time of the motion of a body is expressed by the equation $s = f(t)$, $\frac{ds}{dt}$ is the velocity at any time t. If the velocity has different values for different values of t, represent the velocity at time t_0 by v_0, at time t by v. The average rate of change of velocity in the interval of time $t - t_0$ is $\frac{v - v_0}{t - t_0}$. Calling the interval of time Δt, the change in velocity Δv, the ratio $\frac{\Delta v}{\Delta t}$ measures the average rate of change of velocity in the interval of time Δt. This is true whatever may be the magnitude of the interval of time Δt. The limit of this ratio when limit $\Delta t = 0$ is the actual rate of change of velocity at time t, for the interval of time during which the rate of change of velocity might vary continually decreases. The actual rate of change of velocity is called acceleration. Hence acceleration $= \text{limit } \frac{\Delta v}{\Delta t} = \frac{dv}{dt} = \frac{d}{dt}\frac{ds}{dt} = \frac{d^2s}{dt^2}$. For example, if $s = 16.08\, t^2$, denoting the acceleration by a, $a = 32.16$.

As a direct consequence of the definition

$$\frac{d^2}{dx^2} f(x) = \frac{d}{dx}\left[\frac{d}{dx} f(x)\right], \text{ it follows that } \int \frac{d^2}{dx^2} f(x) = \frac{d}{dx} f(x) + C.$$

For example, let it be required to find the function whose second derivative is $2x - 5$. Denoting the function by $f(x)$, $\frac{d^2}{dx^2} f(x) = 2x + 5$. Integrating, $\frac{d}{dx} f(x) = x^2 - 5x + C_1$. Integrating again, $f(x) = \frac{1}{3}x^3 - \frac{5}{2}x^2 + C_1 x + C_2$. C_1 and C_2 are two arbitrary constants of integration, whose values become known if the function and its first derivative are required to take given values for given values of x. For instance, if it is required that $f(x) = 10$ when $x = 2$ and $\frac{d}{dx}f(x) = 1$ when $x = 0$, $C_1 = 1$ and $C_2 = 15\frac{1}{3}$. Under these conditions, if

$$\frac{d^2}{dx^2}f(x) = 2x - 5, \quad f(x) = \frac{1}{3}x^3 - \frac{5}{2}x^2 + x + 15\frac{1}{3}.$$

PROBLEMS

Find the first and second derivatives of,

1. $3x^2$.
2. $4x^2 - 5$.
3. $2x^3 - 7x$.
4. $\dfrac{1}{x}$.
5. $\dfrac{1}{1-x}$.
6. $\dfrac{x}{1+x^2}$.

Find between what values of x the curves which represent the following equations are convex upward, concave upward, and find the points of inflection,

7. $y = x^3 - 7x + 7$.
8. $y = \dfrac{x}{1+x^2}$.
9. $y = \dfrac{x^3}{3} - x^2 + 2$.

Determine the first and second derivatives of y with respect to x in the implicit functions,

10. $x^2 + y^2 = r^2$. The first derivative is $\dfrac{dy}{dx} = -\dfrac{x}{y}$.

Differentiating both sides of this equation with respect to x,

$$\frac{d^2y}{dx^2} = -\frac{y - x\dfrac{dy}{dx}}{y^2}.$$

Substituting $\dfrac{dy}{dx} = -\dfrac{x}{y}$, there results $\dfrac{d^2y}{dx^2} = -\dfrac{r^2}{y^3}$.

11. $\dfrac{x^2}{a^2} + \dfrac{y^2}{b^2} = 1$.
12. $\dfrac{x^2}{a^2} - \dfrac{y^2}{b^2} = 1$.
13. $y^2 = 2px$.

Find $\dfrac{d^2y}{dx^2}$ from the following equations,

14. $\left(\dfrac{dy}{dx}\right)^2 = y^3$. Differentiating both sides of this equation with respect to x, $2\dfrac{dy}{dx}\dfrac{d^2y}{dx^2} = 3y^2\dfrac{dy}{dx}$, whence $\dfrac{d^2y}{dx^2} = \tfrac{3}{2}y^2$.

15. $\dfrac{dy}{dx} = y^2$.

16. $\dfrac{dy}{dx} = xy$.

17. $\dfrac{dy}{dx} = \dfrac{y}{x}$.

18. $\left(\dfrac{dy}{dx}\right)^2 = y$.

19. $\left(\dfrac{dy}{dx}\right)^2 = \tfrac{2}{3}y^3$.

20. The equation $f(x, y) = 0$ defines y as a continuous function of x. Consequently it also defines x as a continuous function of y. It has been shown that $\dfrac{dy}{dx} = \dfrac{1}{\dfrac{dx}{dy}}$. Prove that

$$\dfrac{d^2y}{dx^2} = -\dfrac{\dfrac{d^2x}{dy^2}}{\left(\dfrac{dx}{dy}\right)^3}.$$

A body moves in such a manner that the relation between s and t is expressed by the following equations. Find velocity and acceleration at time t. a and u are constants.

21. $s = u \cdot t + \tfrac{1}{2}a \cdot t^2$.

22. $s = u \cdot t - \tfrac{1}{2}a \cdot t^2$.

23. $s = u \cdot t^2 + a \cdot t^3$.

24. $s = u \cdot t^2 - a \cdot t^3$.

Determine the functions whose second derivatives are,

25. $3x + 2$, knowing that when $x = 1$, $f(x) = 10$; when $x = 0$, $\dfrac{d}{dx}f(x) = 3$.

26. $5x - 7x^2$, knowing that when
$$x = 0, \; f(x) = 0, \; \dfrac{d}{dx}f(x) = 1.$$

27. Find the curve through the origin and making an angle of $45°$ with the X-axis at the origin, knowing that $\dfrac{d^2y}{dx^2} = 2x + 5$.

28. Find the curve through $(5, 7)$ and parallel to the X-axis at $x = 3$, knowing that $\dfrac{d^2y}{dx^2} = 2x^2 - 4x$.

29. A body starts moving with a velocity u and has a uniform acceleration a. Find the relation between s and t. The conditions of the problem are expressed by $\dfrac{d^2s}{dt^2} = a$, and when $t = 0$, $s = 0$, $\dfrac{ds}{dt} = u$. By integration

$$\frac{ds}{dt} = a \cdot t + C_1, \quad s = \tfrac{1}{2} a \cdot t^2 + C_1 \cdot t + C_2.$$

Since $s = 0$ and $\dfrac{ds}{dt} = u$ when $t = 0$, $C_1 = u$ and $C_2 = 0$. Hence the result is $s = ut + \tfrac{1}{2} at^2$.

30. A body starts moving with a velocity u and has a uniform acceleration $-a$. Find the relation between s and t.

31. A horizontal beam has one end fixed. The bending of the beam due to a weight P causes an elongation of the upper surface of the beam and a compression of the lower surface. Some intermediate surface remains unchanged in length and is called the neutral surface. The figure shows a vertical longitudinal section of the upper, lower, and neutral surfaces of the beam. This section of the neutral surface is called the elastic curve. In text-books on Strength of Materials it is proved that for any point (x, y) in the elastic curve, $EI \dfrac{d^2y}{dx^2} = - P \cdot x$, where E is a constant depending on the material of the beam, I a constant depending on the cross-section of the beam. By the nature of the problem $y = 0$ when $x = 0$,

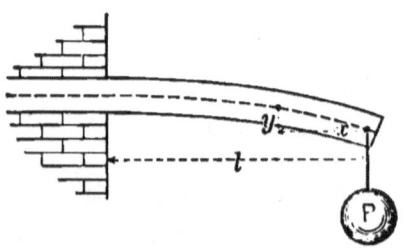

Fig. 18.

and $\frac{dy}{dx}=0$ when $x=l$. Determine the equation of the elastic curve and the maximum deflection of the beam.

32. Suppose the load uniformly distributed over the beam of Problem 31. If the load per linear unit is w and the origin of coordinates is taken at the free end of the beam, it is proved in Strength of Materials that for any point (x, y) in the elastic curve $EI\dfrac{d^2y}{dx^2} = \dfrac{-w \cdot x^2}{2}$. By the nature of the problem $\dfrac{dy}{dx}=0$ when $x=l$, and $y=\Delta$ when $x=l$, where Δ represents the deflection of the free end. Determine the equation of the elastic curve.

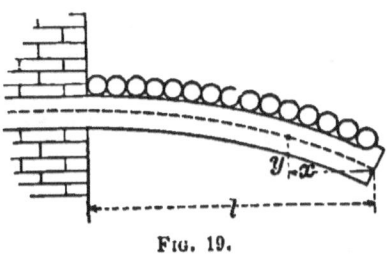
Fig. 19.

33. A horizontal beam rests on two supports the distance between which is l. If a load P is placed at the middle of the beam and the origin of coordinates is taken at the left support, it is proved in Strength of Materials that for any point (x, y) in the elastic curve $EI\dfrac{d^2y}{dx^2} = \dfrac{Px}{2}$. By the nature of the problem $\dfrac{dy}{dx}=0$ when $x=\frac{1}{2}l$, and $y=0$ when $x=0$. Find the equation of the elastic curve and the maximum deflection.

34. The beam of Problem 33 supports a uniform load. If w is the load per linear unit, it is proved in strength of materials that for all points in the elastic curve

$$EI\frac{d^2y}{dx^2} = \frac{wlx}{2} - \frac{wx^2}{2}.$$

By the nature of the problem $\dfrac{dy}{dx}=0$ when $x=\frac{1}{2}l$, and $y=0$

when $x=0$. Find the equation of the elastic curve and the maximum deflection.

35. A circular disc whose weight is W is free to turn about a horizontal axis through the center of the disc and perpendicular to the plane of the disc. A cord wound around the circumference of the disc has a weight P attached to its free end. In Mechanics it is proved that

$$\frac{d^2\theta}{dt^2} = \frac{P \cdot r \cdot g}{W \cdot k^2 + P \cdot r^2},$$

where g is the earth's gravity constant, r is the radius of the disc in feet, k is a constant depending on the size of the disc, and θ is the angle in radians through which the disc turns in t seconds. Find the relation between θ and t.

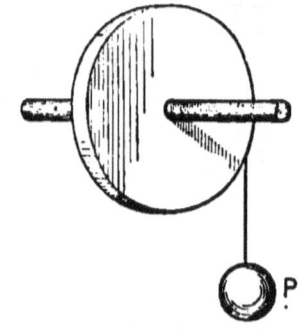

Fig. 20.

In the following problems find $\frac{dy}{dx}$.

36. $\frac{d^2y}{dx^2} = y^3$. Multiply both sides of the given equation by $\frac{dy}{dx}$ and integrate. There results

$$\frac{dy}{dx} \cdot \frac{d^2y}{dx^2} = y^3 \cdot \frac{dy}{dx} \quad \text{and} \quad \tfrac{1}{2}\left(\frac{dy}{dx}\right)^2 = \tfrac{1}{4}y^4 + C.$$

37. $\frac{d^2y}{dx^2} = 2\frac{dy}{dx} + x^2$. **38.** $\frac{d^2y}{dx^2} = 2y^2 - 5y$.

Art. 24. — Maxima and Minima

A continuous function $f(x)$ increases as x increases when its first derivative $\frac{d}{dx}f(x)$ is positive, and decreases as x increases when its first derivative is negative. If $f(x)$ changes from an increasing to a decreasing function when $x = x_0$, the value of $f(x)$ is greater when $x = x_0$ than it is just before x reaches x_0, and also greater than it is just after x passes x_0. This is expressed by saying that $f(x)$ has a maximum value when $x = x_0$. At a maximum value of $f(x)$ therefore the first derivative $\frac{d}{dx} f(x)$ changes sign from $+$ to $-$ for increasing values of x.

If $f(x)$ changes from a decreasing to an increasing function when $x = x_0$, the value of $f(x)$ is less when $x = x_0$ than it is just before x reaches x_0 and also less than it is just after x passes x_0. This is expressed by saying that $f(x)$ has a minimum value when $x = x_0$. If $f(x)$ changes from a decreasing to an increasing function, the first derivative must change sign from $-$ to $+$. Hence at a minimum value of $f(x)$ the first derivative $\frac{d}{dx} f(x)$ changes sign from $-$ to $+$ for increasing values of x.

An algebraic function can change sign only by passing through zero or by passing through infinity. Hence when $f(x)$ has a maximum or a minimum value, $\frac{d}{dx} f(x) = 0$ or $\frac{d}{dx} f(x) = \infty$.

If $\frac{d}{dx} f(x) = 0$ and $f(x)$ has a maximum value, $\frac{d}{dx} f(x)$ changes sign from $+$ to $-$ by passing through zero. Hence $\frac{d}{dx} f(x)$ is a decreasing function, and its first derivative, which is the second derivative of $f(x)$, that is $\frac{d^2}{dx^2} f(x)$, must be negative.

If $\frac{d}{dx} f(x) = 0$ and $f(x)$ has a minimum value, $\frac{d}{dx} f(x)$ changes sign from $-$ to $+$ by passing through zero; hence $\frac{d}{dx} f(x)$ is an increasing function, and its first derivative, which is the second derivative of $f(x)$, must be positive.

These results, obtained analytically, may also be directly

obtained from the graph of $y = f(x)$. For increasing values of x the point generating the continuous curve which represents the equation $y = f(x)$ can cease moving away from the X-axis and start moving towards the X-axis only in one of two ways, either by tending to move parallel to the X-axis, as at P_1,

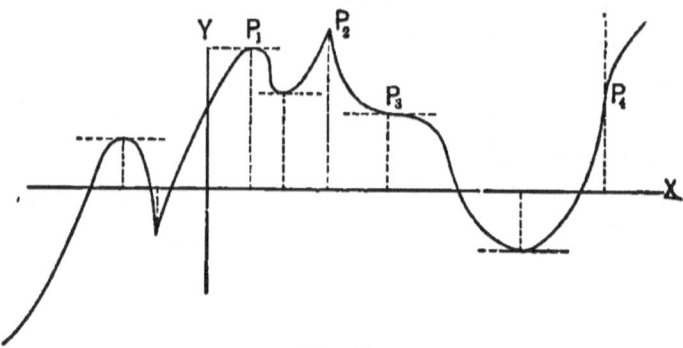

Fig. 21.

which requires that $\frac{dy}{dx} = 0$, or by tending to move parallel to the Y-axis, as at P_2, which requires that $\frac{dy}{dx} = \infty$. At P_3, while $\frac{dy}{dx} = 0$, y is neither a maximum nor a minimum. At P_4, while $\frac{dy}{dx} = \infty$, y is neither a maximum nor a minimum. An examination of the figure shows that in all cases for increasing values of x the slope of the tangent, that is $\frac{dy}{dx}$, changes sign from $+$ to $-$ at a maximum, from $-$ to $+$ at a minimum. At a maximum, when $\frac{dy}{dx} = 0$, the curve is convex upward, and hence $\frac{d^2y}{dx^2}$ is negative; at a minimum, when $\frac{dy}{dx} = 0$, the curve is concave upward, and $\frac{d^2y}{dx^2}$ is positive. From the figure it is evident that maximum and minimum values must occur alternately and that a minimum may be greater than a maximum.

From this investigation is inferred the following method of examining a function of one variable $f(x)$ for maximum and minimum values. Find the roots of the equations $\frac{d}{dx}f(x) = 0$ and $\frac{d}{dx}f(x) = \infty$. If, for increasing values of x, $\frac{d}{dx}f(x)$ changes sign from $+$ to $-$ when x passes one of these roots, this root makes $f(x)$ a maximum, if $\frac{d}{dx}f(x)$ changes sign from $-$ to $+$ when x passes one of these roots, this root makes $f(x)$ a minimum; if $\frac{d}{dx}f(x)$ does not change sign when x passes through one of these roots, this root makes $f(x)$ neither a maximum nor a minimum.

The roots of the equation $\frac{df(x)}{dx} = 0$ which make $\frac{d^2f(x)}{dx^2}$ negative make $f(x)$ a maximum; the roots of $\frac{df(x)}{dx} = 0$ which make $\frac{d^2f(x)}{dx^2}$ positive make $f(x)$ a minimum.

EXAMPLE I. — Examine $x^3 - 7x + 7$ for maximum and minimum values. Here $f(x) = x^3 - 7x + 7$, $\frac{df(x)}{dx} = 3x^2 - 7$, $\frac{d^2f(x)}{dx^2} = 6x$. Equating the first derivative to zero, $3x^2 = 7$ and $x = \pm\sqrt{\tfrac{7}{3}}$. $x = +\sqrt{\tfrac{7}{3}}$ makes $\frac{d^2f(x)}{dx^2}$ positive and $f(x) = -.145$, a minimum. $x = -\sqrt{\tfrac{7}{3}}$ makes $\frac{d^2f(x)}{dx^2}$ negative and $f(x) = 14.145$, a maximum.

EXAMPLE II. — Examine $y = \dfrac{(x+2)^3}{(x-3)^2}$ for maxima and minima. Here $\dfrac{dy}{dx} = \dfrac{(x+2)^2(x-13)}{(x-3)^3}$. The first derivative equated to zero gives $x = -2$, $x = +13$; the first derivative equated to infinity gives $x = +3$. When x is just less than -2, the signs of the three factors of $\dfrac{dy}{dx}$ are $\dfrac{+\,-}{-}$ and $\dfrac{dy}{dx}$ is positive; when x is just greater than -2, the signs of the factors of $\dfrac{dy}{dx}$ are $\dfrac{+\,-}{-}$ and $\dfrac{dy}{dx}$ is still positive. Since $\dfrac{dy}{dx}$

does not change sign when x passes through -2, y is neither a maximum nor a minimum when $x = -2$.

When x is just less than $+13$, the signs of the factors of $\frac{dy}{dx}$ are $\frac{+-}{+}$ and $\frac{dy}{dx}$ is negative; when x is just greater than $+13$, the signs of the factors of $\frac{dy}{dx}$ are $\frac{++}{+}$ and $\frac{dy}{dx}$ is positive. Since $\frac{dy}{dx}$ changes sign from $-$ to $+$ when x passes through 13, y is a minimum when $x = 13$. This minimum is $y = 33\frac{3}{4}$.

When x is just less than $+3$, the signs of $\frac{dy}{dx}$ are $\frac{+-}{-}$ and $\frac{dy}{dx}$ is positive; when x is just greater than $+3$, the signs of $\frac{dy}{dx}$ are $\frac{+-}{+}$ and $\frac{dy}{dx}$ is negative. Since $\frac{dy}{dx}$ changes sign from $+$ to $-$ when x passes through $+3$, y is a maximum when $x = +3$. This maximum is $y = \infty$.

PROBLEMS

Examine the following functions for maximum and minimum values:

1. $x^2 - 3x + 5$.
2. $x^3 - 4x + 7$.
3. $2x^3 - 5x$.
4. $x^3 - 2x^2$.
5. $\frac{1}{4}x^4 - \frac{2}{3}x^3 - \frac{1}{2}x^2 + 2x$.
6. $x^3 - 5x^2 - 10x + 4$.
7. $\frac{(x-1)^2}{(x+1)^3}$.
8. $\frac{x}{1+x^2}$.
9. $\frac{x-1}{(x+2)^2}$.
10. $(x-1)^4(x+2)^3$.
11. $y = 2x - x^2$.
12. $y = 2Rx - x^2$.
13. $y = \frac{b^2}{a^2}(2ax - x^2)$.
14. $y = x^3 + 3x - 5$.
15. $y = \frac{3x-5}{(2x-3)^2}$.

16. $y = 10\sqrt{8x - x^2}$. Suggestion. $10\sqrt{8x - x^2}$ is a maximum when $8x - x^2$ is a maximum, a minimum when $8x - x^2$ is a minimum. Hence constant factors may be dropped and the radical sign removed before forming the first derivative.

17. $\sqrt{3x^2 - 10x + 6}$.   **18.** $\sqrt{\dfrac{x-1}{x+3}}$.   **19.** $3\sqrt{5x^3 - 10x^2}$.

20. Show that $x^3 - 3x^2 + 6x$ has neither a maximum nor a minimum value.

21. Show that $\dfrac{ac - b^2}{a}$ is a maximum value of $ax^2 + 2bx + c$ when a is negative, a minimum value when a is positive.

22. Divide a into two parts whose sum is a minimum.

23. Divide a into two parts such that the sum of their squares is a minimum.

24. Divide a into two parts such that their product is a maximum.

25. Divide a into two parts such that the sum of their square roots is a maximum.

26. From a square sheet of tin 18 inches on a side equal squares are cut at the four corners. From the remainder of the sheet of tin a vessel with open top is formed by bending up the sides. Find side of small squares when the vessel holds the greatest quantity of water.

27. From a rectangular sheet of tin 3 feet by 2 feet equal squares are cut at the four corners and a box with open top formed by turning up the sides. Find sides of squares cut off when contents of box are greatest.

28. A box, square base and open top, is to be constructed to contain 108 cubic inches. What must be its dimensions to require the least material?

29. A circular cylindric standpipe is to be built to hold 10,000 cubic feet of water. Find altitude and diameter of base which require least material.

30. Find the shortest distance from (3, 5) to the line $\dfrac{x}{2} - \dfrac{y}{3} = 1$.

31. Find the shortest distance from (7, 8) to the parabola $y^2 = 4x$.

32. Find the shortest line that can be drawn through (a, b) meeting the rectangular axes.

33. A Norman window is composed of a rectangle surmounted by a semicircle. Find the dimensions of the window so that with a given perimeter the window admits the greatest amount of light.

34. Two trains are running, the one due east at 30 miles per hour, the other due north at 40 miles per hour. When the first train is 30 miles from the intersection of the tracks the second is 20 miles from this point. Find the least distance between the trains.

35. A person in a boat 3 miles from the nearest point of the beach wishes to reach in the shortest time a place 5 miles from that point along the beach. If he can walk 5 miles an hour, but row only 4 miles an hour, where must he land?

36. The strength of a rectangular beam varies as the product of the breadth and the square of the depth. What are the dimensions of the strongest beam that can be cut from a log whose cross-section is a circle 18 inches in diameter? Strength prevents breaking.

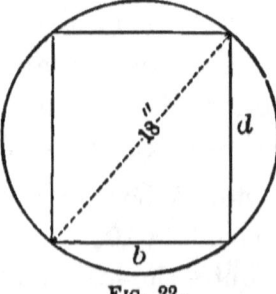

Fig. 22.

37. The stiffness of a rectangular beam varies as the product of the breadth and the cube of the depth. Find the dimensions of the stiffest beam that can be cut from a log whose cross-section is a circle 18 inches in diameter. Stiffness prevents bending.

38. The bending moment of a simple beam whose length is l when the uniform load per linear unit is w is

$$M = \tfrac{1}{2}wlx - \tfrac{1}{2}wx^2,$$

at the point whose distance from the left support is x. Find where the bending moment is a maximum.

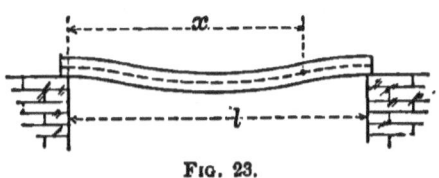

Fig. 23.

39. The distance between two lights A and B is d. The intensity of A at unit's distance is a, of B is b. If the intensity of a light varies inversely as the square of the distance, find the points between the lights of maximum and minimum illumination.

40. If the illumination varies as the sine of the angle under which light strikes the illuminated surface divided by the square of the distance from the source of light, find the height of an electric light directly over the center of a circle of radius r when the illumination of the circumference is greatest.

41. If $c^2 r + \dfrac{t^2}{r}$ is the total waste per mile going on in an electric conductor, r resistance in ohms per mile of conductor, c the current in amperes, t a constant depending on interest on investment and depreciation of plant, find the relation between resistance and current when the waste is a minimum.

42. If v is the velocity of an ocean current in knots per hour, x the velocity of a ship in still water, and if the quantity of fuel burnt per hour is proportional to x^3, find the value of x

which makes the consumption of fuel a minimum for a run of s miles.

43. Given n voltaic cells of E. M. F. e and internal resistance r, to find the way in which the cells should be arranged to send a maximum current through a given external resistance R.

Let x cells be placed in series, then the current $I = \dfrac{xe}{\dfrac{x^2 r}{n} + R}$.

44. The equation of the path of a projectile is

$$y = x \tan \theta - \frac{gx^2}{2\,u^2 \cos^2 \theta},$$

where θ is the angle of projection, u the velocity of projection, $g = 32.16$. The range, the value of x when $y = 0$, is $\dfrac{u^2 \sin(2\theta)}{g}$.

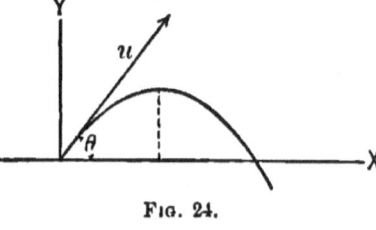

Fig. 24.

a. Find the greatest height.

b. Find the angle of projection which gives the greatest height for a given velocity.

c. Find the angle of projection which gives the greatest range for a given velocity.

45. Find the dimensions of the isosceles triangle of maximum area that can be inscribed in a circle of radius r.

Calling altitude of triangle x, base $2y$, $y^2 = 2rx - x^2$, and area $A = x\sqrt{2rx - x^2}$. A is a maximum when $A' = 2rx^3 - x^4$ is a maximum.

$$\frac{dA'}{dx} = 6rx^2 - 4x^3,$$

which is zero when $x = \tfrac{3}{2}r$. The first derivative changes sign from $+$ to $-$ when x passes through $\tfrac{3}{2}r$. Therefore

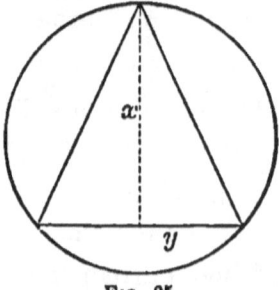

Fig. 25.

the area of the inscribed triangle is a maximum when its altitude is $\frac{3}{2}r$.

46. Find the dimensions of the rectangle of maximum area that can be inscribed in the ellipse $\frac{x^2}{a^2} + \frac{y^2}{b^2} = 1$. The area of

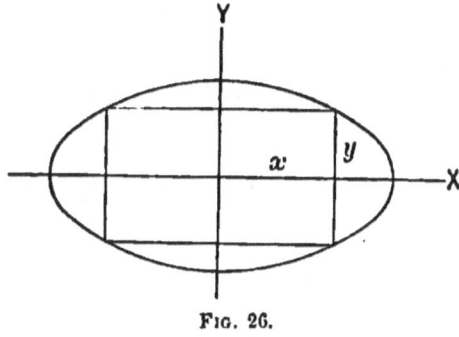

Fig. 26.

the rectangle is $A = 4xy$, which is a maximum when $A' = xy$ is a maximum.

$$\frac{dA'}{dx} = x\frac{dy}{dx} + y.$$

From the equation of the ellipse $\frac{dy}{dx} = -\frac{b^2 x}{a^2 y}$. By substitution, $\frac{dA'}{dx} = \frac{a^2 y^2 - b^2 x^2}{a^2 y}$. Hence A' is a maximum when $a^2 y^2 - b^2 x^2 = 0$. Combining this equation with the equation of the ellipse, $x = \frac{a}{\sqrt{2}}$, $y = \frac{b}{\sqrt{2}}$, and the area of the maximum rectangle is $2ab$.

47. Find the area of the maximum rectangle that can be inscribed in the parabola $y^2 = 2px$ whose limiting coordinates are a, b.

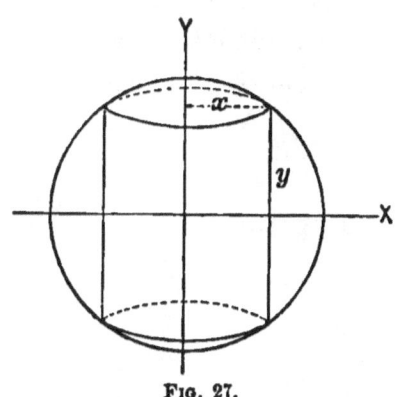

Fig. 27.

48. Find the dimensions of the cylinder of revolution of maximum volume that can be inscribed in a sphere of radius r.

Calling the radius of the base of the cylinder x, the altitude $2y$, (x, y) is a point of the circle $x^2 + y^2 = r^2$ which generates the sphere. Calling the volume of the cylinder V, $V = 2\pi y x^2 = 2\pi y (r^2 - y^2)$. V is a maximum when $\dfrac{dV}{dy} = 2\pi r^2 - 6\pi y^2 = 0$, that is when $y = \dfrac{r}{\sqrt{3}}$, since this value of y makes $\dfrac{d^2 V}{dy^2} = -12\pi y$ negative. The maximum is $V = \dfrac{4}{3\sqrt{3}} \pi r^3$. The ratio of the volume of the maximum inscribed cylinder to the volume of the sphere is $\dfrac{1}{\sqrt{3}}$.

49. Find the cone of maximum volume that can be inscribed in a sphere of radius r.

50. Find the cylinder of maximum volume that can be inscribed in a cone, altitude h, radius of base r.

51. Find the maximum cylinder that can be inscribed in the prolate spheroid.

52. Find the maximum cylinder that can be inscribed in the oblate spheroid.

53. Find the cone of maximum volume that can be inscribed in a paraboloid of revolution, the limiting point of the generating parabola $y^2 = 2px$ being (a, b), the vertex of the cone at intersection of axis of paraboloid with base of paraboloid.

54. Of all right circular cones whose convex surface is the same find the dimensions of that whose volume is greatest.
$V = \tfrac{1}{3}\pi y^2 x$, $\dfrac{dV}{dx} = \tfrac{1}{3}\pi\left(2xy\dfrac{dy}{dx} + x^2\right)$ and $\pi y \sqrt{x^2 + y^2} =$ constant.
$x^2 y^2 + y^4 =$ constant, and $\dfrac{dy}{dx} = \dfrac{-xy}{x^2 + 2y^2}$.
Hence $\dfrac{dV}{dx} = \tfrac{1}{3}\pi\left(\dfrac{2y^4 - x^2 y^2}{x^2 + 2y^2}\right)$. Equating the first derivative to zero, $x = y\sqrt{2}$, which makes V a maximum since $\dfrac{dV}{dx}$ changes sign from $+$ to $-$ when x passes through $y\sqrt{2}$.

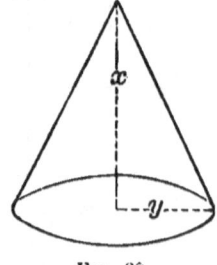

Fig. 28.

74   DIFFERENTIAL AND INTEGRAL CALCULUS

55. Of all right circular cylinders whose convex surface plus the surface of the lower base is constant, find the dimensions of that whose volume is greatest.

56. The distance between the centers of two spheres of radii r and R is d. Find on the line joining the centers the point from which the greatest amount of spherical surface is visible.

57. Find the axis of the parabola of maximum area that can be cut from a right circular cone, radius of base r, altitude h.

ART. 25. — DERIVATIVES OF HIGHER ORDERS

It has been found convenient to call the first derivative of the first derivative of $f(x)$ the second derivative of $f(x)$, and to denote it by $\dfrac{d^2}{dx^2} f(x)$ or $f''(x)$. In like manner it is found convenient to call the first derivative of the second derivative the third derivative of $f(x)$, and to denote it by the symbol $\dfrac{d^3}{dx^3} f(x)$ or $f'''(x)$. By an extension of the same notation $f^{\text{iv}}(x)$, $f^{\text{v}}(x)$, $\cdots$, $f^n(x)$ denote the fourth, fifth, $\cdots$, nth derivatives of $f(x)$.

For example, if $f(x)\ \ = x^4 + 3\,x^3 - 7\,x^2 - 27\,x - 18$,

$$f'(x) = 4\,x^3 + 9\,x^2 - 14\,x - 27,$$
$$f''(x) = 12\,x^2 + 18\,x - 14,$$
$$f'''(x) = 24\,x + 18,$$
$$f^{\text{iv}}(x) = 24,$$
$$f^{\text{v}}(x) = 0.$$

It follows immediately from the definition that the integral of any derivative is the next lower derivative.

For example, if

$f^{IV}(x) = 24x - 18$,
$f'''(x) = 12x^2 - 18x + C_1$,
$f''(x) = 4x^3 - 9x^2 + C_1 \cdot x + C_2$,
$f'(x) = x^4 - 3x^3 + \frac{1}{2}C_1 \cdot x^2 + C_2 \cdot x + C_3$,
$f(x) = \frac{1}{5}x^5 - \frac{3}{4}x^4 + \frac{1}{6}C_1 \cdot x^3 + \frac{1}{2}C_2 \cdot x^2 + C_3 \cdot x + C_4$.

The arbitrary constants C_1, C_2, C_3, C_4 become known if the function $f(x)$ and its first three derivatives $f'(x)$, $f''(x)$, $f'''(x)$ are required to take given values for given values of x.

The successive derivatives $f'(x)$, $f''(x)$, $f'''(x)$, $\cdots$, are called derivatives of the first, second, third, $\cdots$, orders. If $f^n(x)$ is given, $f(x)$ is found by n successive integrations, and the general expression for $f(x)$ must therefore contain n arbitrary constants.*

PROBLEMS

Form the derivatives of the first three orders of,

1. $x^5 - 6x^2 + 15$.   2. $3x^2 + 4x - 7$.   3. $\dfrac{1}{1-x}$.

Find $f(x)$ when,

4. $f'''(x) = x^2 - 5x$.   6. $f'''(x) = 15x^2 + 7$.
5. $f^{IV}(x) = 1 + x$.   7. $f'''(x) = 10$.

ART. 26. — EVALUATION OF THE INDETERMINATE FORM $\dfrac{0}{0}$

The ratio $\dfrac{0}{0}$ may have any value whatever; that is, the value of the ratio is indeterminate. If, however, the ratio of

---

* In Lagrange's notation the successive derivatives of $f(x)$ are denoted by $f'(x)$, $f''(x)$, $f'''(x) \cdots$; in Cauchy's notation, by $Df(x)$, $D^2f(x)$, $D^3f(x) \cdots$. Newton denotes $\dfrac{d^2s}{dt^2}$ by $\ddot{s}$.

two functions $\frac{f(x)}{\phi(x)}$ takes the form $\frac{0}{0}$ for some special value of x, such as $x = a$, the true value of $\frac{f(x)}{\phi(x)}$ when $x = a$ is defined as the limit of the ratio $\frac{f(a + \Delta x)}{\phi(a + \Delta x)}$ when limit $\Delta x = 0$. Since by hypothesis $f(a) = 0$ and $\phi(a) = 0$,

$$\text{limit } \frac{f(a + \Delta x)}{\phi(a + \Delta x)} = \text{limit } \frac{\frac{f(a + \Delta x) - f(a)}{\Delta x}}{\frac{\phi(a + \Delta x) - \phi(a)}{\Delta x}} = \frac{f'(a)}{\phi'(a)},$$

when limit $\Delta x = 0$. Hence by the definition the true value of $\frac{f(x)}{\phi(x)}$ when $x = a$ is $\frac{f'(a)}{\phi'(a)}$, where $f'(a)$ and $\phi'(a)$ denote the values of $f'(x)$ and $\phi'(x)$ when a is substituted for x.

If $\frac{f'(a)}{\phi'(a)} = \frac{0}{0}$, the same analysis shows that the true value of $\frac{f(a)}{\phi(a)} = \frac{f'(a)}{\phi'(a)} = \frac{f''(a)}{\phi''(a)}$. In general, the true value of $\frac{f(a)}{\phi(a)}$ is the first determinate ratio $\frac{f^n(a)}{\phi^n(a)}$ of derivatives of the same order.

Ex. — Find the true value of $\frac{x^3 - 3x + 2}{x^3 - x^2 - x + 1}$ when $x = 1$.

Here $\frac{f(x)}{\phi(x)} = \frac{x^3 - 3x + 2}{x^3 - x^2 - x + 1} = \frac{0}{0}$, when $x = 1$;

$\frac{f'(x)}{\phi'(x)} = \frac{3x^2 - 3}{3x^2 - 2x - 1} = \frac{0}{0}$, when $x = 1$;

$\frac{f''(x)}{\phi''(x)} = \frac{6x}{6x - 2} = \frac{3}{2}$, when $x = 1$.

Hence $\frac{3}{2}$ is the true value of $\frac{x^3 - 3x + 2}{x^3 - x^2 - x + 1}$ when $x = 1$.

PROBLEMS

Find the true values of,

1. $\dfrac{x^3 - 7x^2 + 4x + 12}{x^2 - 14x + 24}$ when $x = 2$.

2. $\dfrac{x^2 - 16}{x^2 + x - 20}$ when $x = 4$.

3. $\dfrac{x^5 - 1}{x^2 - 1}$ when $x = 1$.

4. $\dfrac{5x^2 - 8x + 3}{7x^2 - 9x + 2}$ when $x = 1$.

5. $\dfrac{x^4 - 2x^3 + 2x - 1}{x^5 - 15x^2 + 24x - 10}$ when $x = 1$.

6. $\dfrac{x^4 + 3x^3 - 7x^2 - 27x - 18}{x^4 - 3x^3 - 7x^2 + 27x - 18}$ when $x = 3$.

7. $\dfrac{x^n - 1}{x - 1}$ when $x = 1$.

8. $\dfrac{x^n - a^n}{x - a}$ when $x = a$.

9. $\dfrac{x - 2}{(x - 1)^2 - 1}$ when $x = 2$.

10. $\dfrac{\sqrt{x} - \sqrt{a} + \sqrt{x - a}}{\sqrt{x^2 - a^2}}$ when $x = a$.

11. $\dfrac{(a^2 - x^2)^{\frac{3}{2}}}{(a - x)^{\frac{3}{2}}}$ when $x = a$. Here $\dfrac{f^n(x)}{\phi^n(x)} = \dfrac{0}{0}$ when $x = a$ for all values of n and the true value of the ratio cannot be found by the method of derivatives. The removal of the factor $(a - x)^{\frac{3}{2}}$ from numerator and denominator leads to a determinate result.

CHAPTER VI

PARTIAL DIFFERENTIATION AND INTEGRATION OF ALGEBRAIC FUNCTIONS

Art. 27. — Partial Differentiation

If to pairs of arbitrarily assigned values of x and y there correspond one or more determinate values of z, z is called a function of the two independent variables x and y. This is denoted by writing $z = f(x, y)$.

The function $z = f(x, y)$ is said to be continuous at x_0, y_0 if limit $f(x_0 + \delta_1, y_0 + \delta_2) - f(x_0, y_0) = 0$ when limit $\delta_1 = 0$ and limit $\delta_2 = 0$. It follows that if $f(x, y)$ is a continuous function of x and y, it is also a continuous function of x and y separately. The converse is not necessarily true.

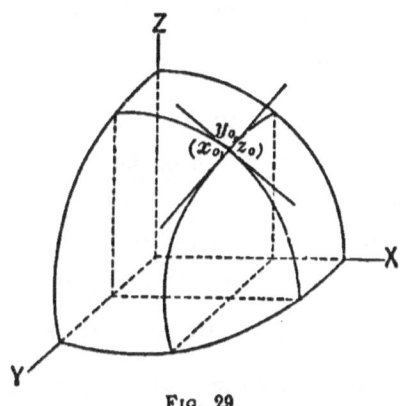

Fig. 29.

The equation $z = f(x, y)$, where x and y are independent variables and z is a continuous function of both x and y, when interpreted in rectangular space coordinates represents a curved surface. Let (x_0, y_0, z_0) be any point in the surface. If x retains the fixed value x_0, the equation $z = f(x_0, y)$ represents the projection on the ZY-plane of the intersection of the plane $x = x_0$ with the surface

78

$z = f(x, y)$. If $z = f(x_0, y)$ is differentiated with respect to y, the result is called the partial derivative of $z = f(x, y)$ with respect to y, and is denoted by the symbol $\dfrac{\partial z}{\partial y}$. Denoting by $\dfrac{\partial z_0}{\partial y_0}$ the value of $\dfrac{\partial z}{\partial y}$ when $x = x_0$, $y = y_0$, $z = z_0$, $\dfrac{\partial z_0}{\partial y_0}$ is the slope of the tangent to $z = f(x_0, y)$ at (y_0, z_0) and measures the rate of change at (x_0, y_0, z_0) of z when the point (x, y, z) moves along the curve of intersection of $x = x_0$ and $z = f(x, y)$.

If y retains the fixed value y_0, the equation $z = f(x, y_0)$ represents the projection on the ZX-plane of the intersection of the plane $y = y_0$ with the surface $z = f(x, y)$. If $z = f(x, y_0)$ is differentiated with respect to x, the result is called the partial derivative of $z = f(x, y)$ with respect to x, and is denoted by the symbol $\dfrac{\partial z}{\partial x}$. Denoting by $\dfrac{\partial z_0}{\partial x_0}$ the value of $\dfrac{\partial z}{\partial x}$ when $x = x_0$, $y = y_0$, $z = z_0$, $\dfrac{\partial z_0}{\partial x_0}$ is the slope of the tangent to $z = f(x, y_0)$ at (z_0, x_0) and measures the rate of change at (x_0, y_0, z_0) of z when the point (x, y, z) moves along the curve of intersection of $y = y_0$ and $z = f(x, y)$.*

The equations of the straight line tangent to the curve of intersection of the plane $x = x_0$ and the surface $z = f(x, y)$ at (x_0, y_0, z_0) are $x = x_0$ and $z - z_0 = \dfrac{\partial z_0}{\partial y_0}(y - y_0)$; the equations of the straight line tangent to the curve of intersection of $y = y_0$ and $z = f(x, y)$ at (x_0, y_0, z_0) are $y = y_0$ and $z - z_0 = \dfrac{\partial z_0}{\partial y_0}(x - x_0)$. The plane containing these two tangent lines is the tangent plane to the surface $z = f(x, y)$ at (x_0, y_0, z_0). In the analytic geometry of three dimensions it is proved that the plane $A(x - x_0) + B(y - y_0) + C(z - z_0) = 0$ through the point (x_0, y_0, z_0) contains the line $x - x_0 = a(z - z_0)$ $y - y_0 = b(z - z_0)$,

---

*The use of ∂ to denote partial differentiation was introduced by Jacobi (1801-1851).

when $Aa + Bb + C = 0$. Hence the equation of the tangent plane is found to be $z - z_0 = \dfrac{\partial z_0}{\partial x_0}(x - x_0) + \dfrac{\partial z_0}{\partial y_0}(y - y_0)$.

Denoting by Z the angle made by the tangent plane with the XY-plane, $\cos Z = \dfrac{1}{\left(1 + \dfrac{\partial z_0^2}{\partial x_0^2} + \dfrac{\partial z_0^2}{\partial y_0^2}\right)^{\frac{1}{2}}}$.

The normal to the curved surface $z = f(x, y)$ at (x_0, y_0, z_0) is the line through (x_0, y_0, z_0) perpendicular to the tangent plane at this point. Hence the equations of the normal are

$$x - x_0 = -\dfrac{\partial z_0}{\partial x_0}(z - z_0), \quad y - y_0 = -\dfrac{\partial z_0}{\partial y_0}(z - z_0).$$

For example, let it be required to find the equation of the tangent plane and the equations of the normal to the sphere $x^2 + y^2 + z^2 = 14$ at $(1, 2, 3)$. Differentiating, regarding y constant, $x + z\dfrac{\partial z}{\partial x} = 0$. Differentiating, regarding x constant, $y + z\dfrac{\partial z}{\partial y} = 0$. At the point of tangency

$$x_0 = 1, \ y_0 = 2, \ z_0 = 3, \ \dfrac{\partial z_0}{\partial x_0} = -\dfrac{1}{3}, \ \dfrac{\partial z_0}{\partial y_0} = -\dfrac{2}{3}.$$

Hence the equation of the tangent plane is

$$z - 3 = -\tfrac{1}{3}(x - 1) - \tfrac{2}{3}(y - 2),$$

reducing to $x + 2y + 3z = 14$; the equations of the normal are $x - 1 = \tfrac{1}{3}(z - 3)$, $y - 2 = \tfrac{2}{3}(z - 3)$, reducing to $x = \tfrac{1}{3}z$, $y = \tfrac{2}{3}z$.

If the altitude of a right circular cone is y and the radius of its base is x, denoting the volume by V, $V = \tfrac{1}{3}\pi x^2 \cdot y$. The variables x and y are independent, for a change in x does not cause a change in y. Suppose the radius to remain unchanged while the altitude varies. Differentiating partially with respect to y, $\dfrac{\partial V}{\partial y} = \tfrac{1}{3}\pi x^2$. This shows that if the base of the cone remains constant the volume changes $\tfrac{1}{3}\pi x^2$ times as fast as the altitude. If the altitude remains constant and the base changes, $\dfrac{\partial V}{\partial x} = \tfrac{2}{3}\pi xy$. That is, the volume changes $\tfrac{2}{3}\pi xy$ times as fast as the radius of the base.

PROBLEMS

In the following functions x and y are independent variables. Form $\dfrac{\partial z}{\partial x}$ and $\dfrac{\partial z}{\partial y}$.

1. $z = xy$.
2. $z = x^2 + y^2$.
3. $z = 2xy^2 + 5x^2y + 6x - 8y$.
4. $z = (x^2 + y^2)^{\frac{1}{2}}$.
5. $z = \dfrac{x}{y}$.
6. $z = \dfrac{1+x}{1-y}$.
7. $z = \dfrac{x-y}{x+y}$.
8. $x^2 + y^2 + z^2 = r^2$.
9. $z = ax^2 + by^2$.
10. $\dfrac{x^2}{a^2} + \dfrac{y^2}{b^2} + \dfrac{z^2}{c^2} = 1$.

11. Find the equation of the plane tangent to
$$x^2 - 4y^2 + 2z^2 = 6 \text{ at } (2, -2, 3).$$

12. Find the equation of the normal to $x^2 - 4y^2 + 2z^2 = 6$ at $(2, 2, 3)$.

13. Denote the base of a triangle by x, its altitude by y. Find the rate of change of area when the base remains unchanged.

14. The pressure of a gas on the containing vessel varies directly as the temperature and inversely as the volume. That is, denoting pressure by p, temperature by t, and volume by v, $p = c\dfrac{t}{v}$, where c is a constant. Find the rate of change

of pressure when the volume is constant. Also the rate of change of pressure when the temperature is constant.

ART. 28. — PARTIAL INTEGRATION

EXAMPLE. — Find z by integration when $\dfrac{\partial z}{\partial x} = 3yx + x^3 + y^2$.

Since y is considered constant in forming $\dfrac{\partial z}{\partial x}$, terms of z containing only y and constants have no effect on the value of $\dfrac{\partial z}{\partial x}$. Hence the integration of $\dfrac{\partial z}{\partial x}$, considering y constant, gives all the terms of z which contain x, but it is necessary to add an arbitrary function of y, $f(y)$, to cover the terms of z which do not contain x. This integration gives

$$z = \tfrac{3}{2}x^2 y + \tfrac{1}{4}x^4 + y^2 x + f(y).$$

If the value of z is known for some special value of x, the value of $f(y)$ becomes known and the result is determinate. For example, if $z = 3y + 5$ when $x = 0$, $f(y) = 3y + 5$ and $z = \tfrac{3}{2}x^2 y + \tfrac{1}{4}x^4 + y^2 x + 3y + 5$.

PROBLEMS

Integrate,

1. $\dfrac{\partial z}{\partial x} = y^2 x$.

2. $\dfrac{\partial z}{\partial x} = xy^2 - 3x$.

3. $\dfrac{\partial z}{\partial y} = xy^2 - 3x$.

4. $\dfrac{\partial z}{\partial y} = y^2 - x^2$.

5. $z \dfrac{\partial z}{\partial x} = x + \tfrac{1}{2} y^2$.

6. $\dfrac{\partial z}{\partial x} = 3x^2 - 5y$, knowing that $z = 3y^2 - 7$ when $x = 0$.

7. $z \dfrac{\partial z}{\partial x} = 2y^2 - x$, knowing that $z^2 = 10 + y^1$ when $x = 2$.

Art. 29. — Differentiation of Implicit Functions

Represent by Δx and Δy the corresponding changes in the values of x and y when y is a continuous implicit function of x defined by the equation $f(x, y) = 0$.

By hypothesis $f(x, y) = 0$ and $f(x + \Delta x, y + \Delta y) = 0$. By subtraction $f(x + \Delta x, y + \Delta y) - f(x, y) = 0$, whence, by adding and subtracting $f(x, y + \Delta y)$,

$$[f(x + \Delta x, y + \Delta y) - f(x, y + \Delta y)] + [f(x, y + \Delta y) - f(x, y)] = 0.$$

Dividing the last equation by Δx and then multiplying numerator and denominator of the second term by Δy, there results

(1) $$\frac{f(x + \Delta x, y + \Delta y) - f(x, y + \Delta y)}{\Delta x} + \frac{f(x, y + \Delta y) - f(x, y)}{\Delta y} \frac{\Delta y}{\Delta x} = 0.$$

Now since by hypothesis y is a continuous function of x,

$$\frac{f(x + \Delta x, y + \Delta y) - f(x, y + \Delta y)}{\Delta x} = \frac{f(x + \Delta x, y) - f(x, y)}{\Delta x} \pm \epsilon,$$

where ϵ has zero for its limit when limit $\Delta x = 0$. Hence when limit $\Delta x = 0$, (1) becomes

$$\frac{\partial f(x, y)}{\partial x} + \frac{\partial f(x, y)}{\partial y} \frac{dy}{dx} = 0 \text{ and } \frac{dy}{dx} = -\frac{\dfrac{\partial f(x, y)}{\partial x}}{\dfrac{\partial f(x, y)}{\partial y}}.$$

Representing $f(x, y)$ by u, this result becomes

$$\frac{dy}{dx} = -\frac{\dfrac{\partial u}{\partial x}}{\dfrac{\partial u}{\partial y}}.$$

Denoting by $\dfrac{\partial u_0}{\partial x_0}$ and $\dfrac{\partial u_0}{\partial y_0}$ the values of the partial derivatives $\dfrac{\partial u}{\partial x}$ and $\dfrac{\partial u}{\partial y}$ when $x = x_0$, $y = y_0$, the equation of the tangent to the curve whose equation is $u = f(x, y) = 0$ at the point (x_0, y_0) is $(y - y_0)\dfrac{\partial u_0}{\partial y_0} + (x - x_0)\dfrac{\partial u_0}{\partial x_0} = 0$; the equation of the normal is $(y - y_0)\dfrac{\partial u_0}{\partial x_0} - (x - x_0)\dfrac{\partial u_0}{\partial y_0} = 0$.

Let z be a continuous implicit function of the independent variables x and y defined by the equation $f(x, y, z) = 0$. Denoting by Δx and Δz the corresponding changes in the values of x and z when y remains unchanged,

$$f(x + \Delta x, y, z + \Delta z) - f(x, y, z) = 0,$$

whence

$$\frac{f(x + \Delta x, y, z + \Delta z) - f(x, y, z + \Delta z)}{\Delta x}$$

$$+ \frac{f(x, y, z + \Delta z) - f(x, y, z)}{\Delta z} \frac{\Delta z}{\Delta x} = 0.$$

When limit $\Delta x = 0$, this equation becomes

$$\frac{\partial f(x, y, z)}{\partial x} + \frac{\partial f(x, y, z)}{\partial z}\frac{\partial z}{\partial x} = 0, \text{ whence } \frac{\partial z}{\partial x} = -\frac{\dfrac{\partial f(x, y, z)}{\partial x}}{\dfrac{\partial f(x, y, z)}{\partial z}}.$$

In like manner, denoting by Δy and Δz the corresponding changes in y and z when x remains unchanged, and passing to the limit when limit $\Delta y = 0$,

$$\frac{\partial z}{\partial y} = -\frac{\dfrac{\partial f(x, y, z)}{\partial y}}{\dfrac{\partial f(x, y, z)}{\partial z}}.$$

PARTIAL DIFFERENTIATION

Representing $f(x, y, z)$ by u, these results become

$$\frac{\partial z}{\partial x} = -\frac{\frac{\partial u}{\partial x}}{\frac{\partial u}{\partial z}} \quad \text{and} \quad \frac{\partial z}{\partial y} = -\frac{\frac{\partial u}{\partial y}}{\frac{\partial u}{\partial z}}.$$

Denoting by $\frac{\partial u_0}{\partial x_0}, \frac{\partial u_0}{\partial y_0}$, and $\frac{\partial u_0}{\partial z_0}$ the values of the partial derivatives $\frac{\partial u}{\partial x}, \frac{\partial u}{\partial y}, \frac{\partial u}{\partial z}$ when $x = x_0$, $y = y_0$, $z = z_0$, the equation of the tangent plane to the surface whose equation is $u = f(x, y, z) = 0$ at the point (x_0, y_0, z_0) is

$$(x - x_0)\frac{\partial u_0}{\partial x_0} + (y - y_0)\frac{\partial u_0}{\partial y_0} + (z - z_0)\frac{\partial u_0}{\partial z_0} = 0;$$

the equations of the normal are

$$(x - x_0)\frac{\partial u_0}{\partial z_0} - (z - z_0)\frac{\partial u_0}{\partial x_0} = 0, \quad (y - y_0)\frac{\partial u_0}{\partial z_0} - (z - z_0)\frac{\partial u_0}{\partial y_0} = 0.$$

Denoting by $\theta_x, \theta_y, \theta_z$ the angles made by the normal with the rectangular coordinate axes X, Y, Z respectively,

$$\cos \theta_x = \frac{\frac{\partial u_0}{\partial x_0}}{\left(\frac{\partial u_0^2}{\partial x_0^2} + \frac{\partial u_0^2}{\partial y_0^2} + \frac{\partial u_0^2}{\partial z_0^2}\right)^{\frac{1}{2}}},$$

with corresponding values for $\cos \theta_y$, $\cos \theta_z$.

EXAMPLE. — Find equations of tangent plane and of normal to $\frac{x^2}{9} + \frac{y^2}{4} + z^2 = 1$ at $(2, 1, \frac{1}{3}\sqrt{11})$.

Writing $u = \frac{x^2}{9} + \frac{y^2}{4} + z^2 - 1 = 0$, $\frac{\partial u}{\partial x} = \frac{2x}{9}$, $\frac{\partial u}{\partial y} = \frac{y}{2}$, $\frac{\partial u}{\partial z} = 2z$. Hence when

$x_0 = 2$, $y_0 = 1$, $z_0 = \frac{1}{3}\sqrt{11}$, $\frac{\partial u_0}{\partial x_0} = \frac{4}{9}$, $\frac{\partial u_0}{\partial y_0} = \frac{1}{2}$, $\frac{\partial u_0}{\partial z_0} = \frac{1}{3}\sqrt{11}$.

The equation of the tangent plane is

$$\tfrac{4}{9}(x-2) + \tfrac{1}{2}(y-1) + \tfrac{1}{3}\sqrt{11}\,(z - \tfrac{1}{6}\sqrt{11}) = 0.$$

The equations of the normal are

$$\tfrac{1}{3}\sqrt{11}\,(x-2) - \tfrac{4}{9}(z - \tfrac{1}{6}\sqrt{11}) = 0$$

and

$$\tfrac{1}{3}\sqrt{11}\,(y-1) - \tfrac{1}{2}(z - \tfrac{1}{6}\sqrt{11}) = 0.$$

For this normal

$$\cos\theta_x = .3233,\ \theta_x = 71°8';\ \cos\theta_y = .3638,\ \theta_y = 68°20';$$

$$\cos\theta_z = .8341,\ \theta_z = 33°29'.$$

PROBLEMS

1. Find equations of tangent and normal to
$$y^3 - 3xy + x^3 = 0 \text{ at } (x_0, y_0).$$

2. Find equation of tangent plane and normal to
$$x^2 + y^2 - z^2 = 0 \text{ at } (3, 4, 5).$$

3. Find angles made by normal to $x^2 + y^2 - z^2 = 0$ at $(2, 0, 4)$ with the coordinate axes.

4. Find equation of tangent plane to $xyz = 8$ at $(2, 2, 2)$.

5. Find angle made by plane tangent to $z = 2x^2 + 4y^2$ at $(2, 1, 12)$ with XY-plane.

Art. 30. — Successive Partial Differentiation and Integration

If $z = f(x, y)$, the partial derivatives of z with respect to x and y, $\dfrac{\partial z}{\partial x}$ and $\dfrac{\partial z}{\partial y}$, are in general functions of x and y. The partial derivative with respect to x of $\dfrac{\partial z}{\partial x}$ is called the second partial derivative of z with respect to x, and is denoted by the symbol $\dfrac{\partial^2 z}{\partial x^2}$. Hence by definition $\dfrac{\partial^2 z}{\partial x^2} = \dfrac{\partial}{\partial x}\dfrac{\partial z}{\partial x}$. The partial derivative with respect to y of $\dfrac{\partial z}{\partial x}$ is called the second partial derivative of z with respect to x and y, and is denoted by the symbol $\dfrac{\partial^2 z}{\partial y \partial x}$. Hence by definition $\dfrac{\partial^2 z}{\partial y \partial x} = \dfrac{\partial}{\partial y}\dfrac{\partial z}{\partial x}$. By a like notation $\dfrac{\partial^2 z}{\partial y^2} = \dfrac{\partial}{\partial y}\dfrac{\partial z}{\partial y}$, $\dfrac{\partial^2 z}{\partial x \partial y} = \dfrac{\partial}{\partial x}\dfrac{\partial z}{\partial y}$. An extension of this notation gives $\dfrac{\partial^3 z}{\partial x^3} = \dfrac{\partial}{\partial x}\dfrac{\partial^2 z}{\partial x^2}$, $\dfrac{\partial^3 z}{\partial y \partial x^2} = \dfrac{\partial}{\partial y}\dfrac{\partial^2 z}{\partial x^2}$, $\dfrac{\partial^3 z}{\partial x \partial y \partial x} = \dfrac{\partial}{\partial x}\dfrac{\partial^2 z}{\partial y \partial x}$, and so on.

For example, if $z = x^3 + 3xy^2 + yx^2$, $\dfrac{\partial z}{\partial x} = 6x^2 + 3y^2 + 2yx$, $\dfrac{\partial^2 z}{\partial x^2} = 12x + 2y$, $\dfrac{\partial^2 z}{\partial y \partial x} = 6y + 2x$, $\dfrac{\partial z}{\partial y} = 6xy + x^2$, $\dfrac{\partial^2 z}{\partial x \partial y} = 6y + 2x$.

In this example $\dfrac{\partial^2 z}{\partial x \partial y} = \dfrac{\partial^2 z}{\partial y \partial x}$. It is to be proved that this is always true. The proposition may be formulated thus:

If the function $z = f(x, y)$ and its partial derivatives $\dfrac{\partial z}{\partial x}$, $\dfrac{\partial z}{\partial y}$, $\dfrac{\partial^2 z}{\partial x \partial y}$, $\dfrac{\partial^2 z}{\partial y \partial x}$ have determinate finite values, and z, $\dfrac{\partial z}{\partial x}$, $\dfrac{\partial z}{\partial y}$ are continuous functions of x and y, $\dfrac{\partial^2 z}{\partial y \partial x} = \dfrac{\partial^2 z}{\partial x \partial y}$.

By hypothesis $\dfrac{\partial z}{\partial x} = \dfrac{f(x + \Delta x, y) - f(x, y)}{\Delta x} \pm \epsilon$, where ϵ depends on x, y, and Δx, and approaches zero with Δx. It is convenient to write $\epsilon = \epsilon_1 f_1(x, y)$, where ϵ_1 depends only on Δx and approaches zero with Δx while $f_1(x, y)$ must be finite. Similarly

$$\dfrac{\partial^2 z}{\partial y \partial x} = \dfrac{f(x+\Delta x, y+\Delta y) - f(x, y+\Delta y) - f(x+\Delta x, y) + f(x, y)}{\Delta x \, \Delta y}$$
$$\pm \epsilon_1 \dfrac{f_1(x, y+\Delta y) - f_1(x, y)}{\Delta y} \pm \epsilon_2 f_2(x, y),$$

where limit $\dfrac{f_1(x, y+\Delta y) - f_1(x, y)}{\Delta y}$, when limit $\Delta y = 0$, must be finite, ϵ_2 must vanish with Δy, and $f_2(x, y)$ must be finite by hypothesis.

In like manner $\dfrac{\partial z}{\partial y} = \dfrac{f(x, y + \Delta y) - f(x, y)}{\Delta y} \pm \epsilon_3 f_3(x, y)$ and

$$\dfrac{\partial^2 z}{\partial x \partial y} = \dfrac{(x + \Delta x, y + \Delta y) - f(x + \Delta x, y) - f(x, y + \Delta y) + f(x, y)}{\Delta y \, \Delta x}$$

$$\pm \epsilon_3 \dfrac{f_3(x + \Delta x, y) - f_3(x, y)}{\Delta x} \pm \epsilon_4 f_4(x, y),$$

where ϵ_3 must vanish with Δy, ϵ_4 must vanish with Δx, limit $\dfrac{f_3(x + \Delta x, y) - f_3(x, y)}{\Delta x}$ when limit $\Delta x = 0$ must be finite, and $f_4(x, y)$ must be finite. It follows that when limit $\Delta x = 0$ and limit $\Delta y = 0$,

$$\text{limit } \dfrac{f(x + \Delta x, y + \Delta y) - f(x + \Delta x, y) - f(x, y + \Delta y) + f(x, y)}{\Delta y \, \Delta x}$$

$$= \dfrac{\partial^2 z}{\partial x \partial y} = \dfrac{\partial^2 z}{\partial y \partial x}.$$

PROBLEMS

Form $\dfrac{\partial z}{\partial x}$, $\dfrac{\partial^2 z}{\partial x^2}$, $\dfrac{\partial^2 z}{\partial y \partial x}$, $\dfrac{\partial z}{\partial y}$, $\dfrac{\partial z^2}{\partial y^2}$, $\dfrac{\partial^2 z}{\partial x \partial y}$ of,

1. $z = x^2 + y^2$. 2. $z = x^2 y$. 3. $z = \dfrac{x+y}{x-y}$. 4. $z = x(y-2)$.

5. $z = \dfrac{x}{x+y}$. 6. $z = x^2 y^2$. 7. $z = x^{\frac{1}{2}} y^{\frac{1}{2}}$. 8. $z = x^3 y^{-\frac{1}{2}}$.

Integrate 9. $\dfrac{\partial^2 z}{\partial x^2} = x^2 y + 3x - 5y + 2$. Integrating, considering y constant, $\dfrac{\partial z}{\partial x} = \tfrac{1}{3} x^3 y + \tfrac{3}{2} x^2 - 5yx + 2x + f_1(y)$.

Integrating again, considering y constant,

$$z = \tfrac{1}{12} x^4 y + \tfrac{1}{2} x^3 - \tfrac{5}{2} yx^2 + x^2 + f_1(y) \cdot x + f_2(y).$$

The arbitrary functions $f_1(y)$ and $f_2(y)$ become known if z and $\dfrac{\partial z}{\partial x}$ are known for some value of x. Suppose that when $x = 0$,

$z = y + 3$, $\dfrac{\partial z}{\partial x} = y^2 + 5$. Then $f_1(y) = y^2 + 5$, $f_2(y) = y + 3$, and the result becomes

$$z = \tfrac{1}{12} x^4 y + \tfrac{1}{2} x^3 + \tfrac{5}{2} y x^2 + x^2 + x y^2 + 5 x + y + 3.$$

10. $\dfrac{\partial^2 z}{\partial x \partial y} = xy + 2x - 7y + 5$, knowing that, when $x = 0$, $z = y - 2$ and when $y = 0$, $\dfrac{\partial z}{\partial x} = 2 x^3$.

11. $\dfrac{\partial^2 z}{\partial y \partial x} = 3 x y^2$.   12. $\dfrac{\partial^2 z}{\partial y^2} = x y^2 - 2$.   13. $\dfrac{\partial^2 z}{\partial x^2} = x^2 - y^2$.

14. $\dfrac{\partial^2 z}{\partial x \partial y} = (x - 3)(y - 2)$.   15. $\dfrac{\partial^2 z}{\partial y \partial x} = x(y - 5)$.

Art. 31. — Area of Any Curved Surface

Let the equation of the curved surface be $z = f(x, y)$ continuous in x and y. Planes perpendicular to the X-axis at intervals of Δx divide the given surface into strips of surface $CDEF$. Planes perpendicular to the Y-axis at intervals of Δy divide these strips into elements of surface $abcd$. The strip $CDEF = \sum\limits_{y=0}^{y=BC} abcd$, and the given surface

$$A = \sum_{x=0}^{x=OA} CDEF = \sum_{x=0}^{x=OA} \sum_{y=0}^{y=BC} abcd.$$

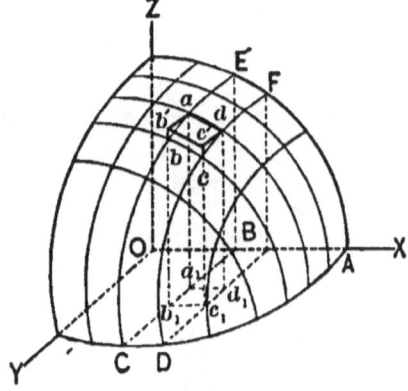

Fig. 80.

This summation holds whatever be the magnitude of Δx and Δy. The projection of the element of surface $abcd$ on the XY-plane is the rectangle $a_1 b_1 c_1 d_1$, whose area is $\Delta x \Delta y$. The plane through ad and the line ab' parallel to $a_1 b_1$ intersects the prism which projects *abcd* on the *XY*-plane in the parallelogram *ab'c'd*. Denoting by θ'_z the angle made by the plane of *ab'c'd* with the *XY*-plane, area $ab'c'd = \dfrac{\Delta x \Delta y}{\cos \theta'_z}$. When Δx and Δy approach zero, the plane of the parallelogram *ab'c'd* approaches the tangent plane to the curved surface at *a*, and the parallelogram *ab'c'd* approaches the surface element *abcd*. Hence, when limit $\Delta x = 0$ and limit $\Delta y = 0$,

$$A = \sum_{x=0}^{x=OA} \sum_{y=0}^{y=BC} abcd = \text{limit} \sum_{x=0}^{x=OA} \sum_{y=0}^{y=BC} ab'c'd = \text{limit} \sum_{x=0}^{x=OA} \sum_{y=0}^{y=BC} \frac{\Delta x \Delta y}{\cos \theta'_z}$$

$$= \int_{x=0}^{x=OA} \int_{y=0}^{y=BC} \left(1 + \frac{\partial z^2}{\partial x^2} + \frac{\partial z^2}{\partial y^2}\right)^{\frac{1}{2}} dx\, dy.$$

EXAMPLE. — Find the area of the groin formed by the intersection of the semicircular cylinders $x^2 + z^2 = r^2$, $y^2 + z^2 = r^2$.

One-eighth of the surface of the groin is bounded by the cylinder $x^2 + z^2 = r^2$, the *ZX*-plane, the *XY*-plane, and the intersection of the cylinders, which intersection lies in the plane $y = x$. Hence for this part of the surface of the groin

$$\frac{\partial z}{\partial x} = -\frac{x}{z}, \quad \frac{\partial z}{\partial y} = 0,$$

and

$$A = 8 \int_{x=0}^{x=r} \int_{y=0}^{y=x} \left(1 + \frac{x^2}{z^2}\right)^{\frac{1}{2}} dx\, dy$$

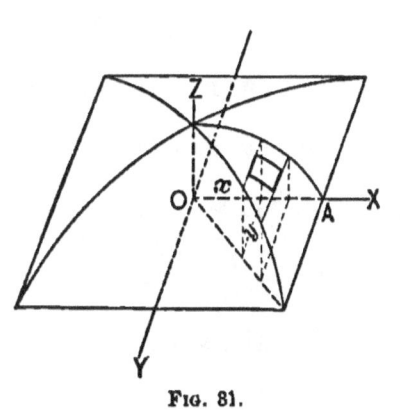

Fig. 81.

$$= 8r \int_{x=0}^{x=r} \int_{y=0}^{y=x} \frac{dx\, dy}{(r^2 - x^2)^{\frac{1}{2}}} = 8r \int_{x=0}^{x=r} \frac{x\, dx}{(r^2 - x^2)^{\frac{1}{2}}}$$

$$= -8r\{(r^2 - x^2)^{\frac{1}{2}} + C\}_0^r = 8r^2.$$

Art. 32. — Volume of Any Solid

Denote by V the volume of the solid bounded by the continuous surface $z = f(x, y)$ and the coordinate planes XY, XZ, YZ. Planes perpendicular to the X-axis at intervals Δx divide the solid into laminæ LL'; planes perpendicular to the Y-axis at intervals Δy divide these laminæ into prisms PP'; planes perpendicular to the Z-axis at intervals Δz divide these prisms into elements of volume aa'. The volume of

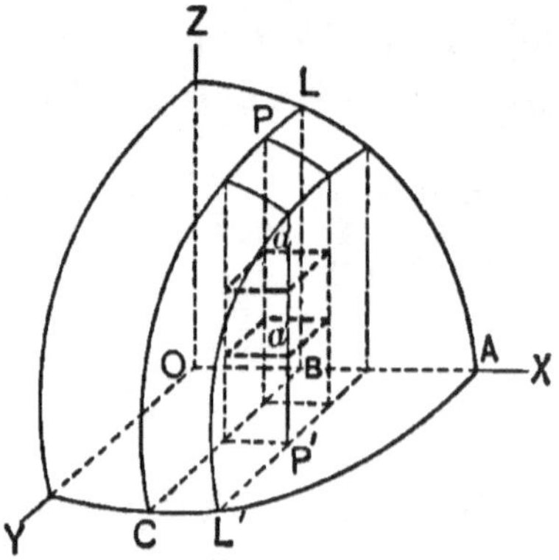

$$aa' = \Delta x \cdot \Delta y \cdot \Delta z;$$

Fig. 32.

the volume of $PP' = \sum\limits_{z=0}^{z=PD} \Delta x \cdot \Delta y \cdot \Delta z - \theta \cdot \Delta x \cdot \Delta y \cdot \Delta z$, where θ is less than unity. When $\Delta x, \Delta y, \Delta z$ approach zero, volume of $PP' = \sum\limits_{z=0}^{z=PD} dx \cdot dy \cdot dz - \theta \cdot dx \cdot dy \cdot dz$. Since $\theta \cdot dx \cdot dy \cdot dz$ is an infinitesimal of a higher order than $\sum\limits_{z=0}^{z=PD} dx \cdot dy \cdot dz = PD \cdot dx \cdot dy$, in the limit of the sum volume of $PP' = \int_{z=0}^{z=PD} dx \cdot dy \cdot dz$. The volume of the lamina $LL' = \sum\limits_{y=0}^{y=BC} PP'$; the volume of the solid is

$$V = \sum_{x=0}^{x=OA} LL' = \sum_{x=0}^{x=OA} \sum_{y=0}^{y=BC} PP'.$$

Hence in the limit

$$V = \int_{x=0}^{x=OA} \int_{y=0}^{y=BC} \int_{z=0}^{z=PD} dx\, dy\, dz.$$

EXAMPLE. — Find the volume of the solid bounded by the ZX-plane, the XY-plane, and the planes $x=a$, $y=b$, $z=mx$. In this problem

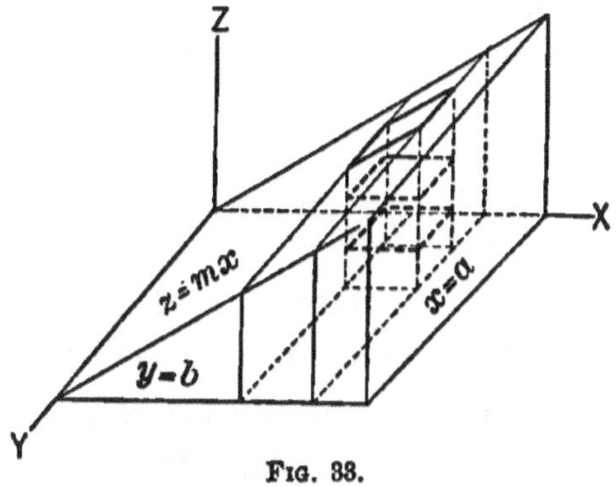

$$V = \int_{x=0}^{x=a} \int_{y=0}^{y=b} \int_{z=0}^{z=mx} dx\,dy\,dz$$

$$= m \int_{x=0}^{x=a} \int_{y=0}^{y=b} x\,dx\,dy$$

$$= mb \int_{x=0}^{x=a} x\,dx = \tfrac{1}{2} mba^2.$$

Fig. 33.

Art. 33. — Total Differentials

Let $u = f(x, y)$ represent a continuous function of the independent variables x and y. Let $\Delta_x u$ denote the change in the value of u corresponding to a change of Δx in the value of x when y remains unchanged; $\Delta_y u$ the change in the value of u corresponding to a change of Δy in the value of y when x remains unchanged; Δu the change in the value of u corresponding to a change of both x and y by Δx and Δy respectively. Then

(1) $$\Delta_x u = \frac{f(x+\Delta x, y) - f(x, y)}{\Delta x} \Delta x,$$

(2) $$\Delta_y u = \frac{f(x, y+\Delta y) - f(x, y)}{\Delta y} \Delta y,$$

(3) $\Delta u = f(x+\Delta x, y+\Delta y) - f(x, y)$

$$= \frac{f(x+\Delta x, y+\Delta y) - f(x, y+\Delta y)}{\Delta x} \Delta x$$

$$+ \frac{f(x, y+\Delta y) - f(x, y)}{\Delta y} \Delta y$$

$$= \frac{f(x+\Delta x, y) - f(x, y) \pm \epsilon}{\Delta x} \Delta x$$

$$+ \frac{f(x, y+\Delta y) - f(x, y)}{\Delta y} \Delta y,$$

where ϵ vanishes when Δy approaches zero. If Δx and Δy are indefinitely decreased, becoming dx and dy, and the corresponding values of $\Delta_x u$, $\Delta_y u$, Δu are denoted by $d_x u$, $d_y u$, du respectively, equations (1), (2), and (3) become

$$d_x u = \frac{\partial u}{\partial x} dx, \quad d_y u = \frac{\partial u}{\partial y} dy, \quad du = \frac{\partial u}{\partial x} dx + \frac{\partial u}{\partial y} dy = d_x u + d_y u.$$

The quantity $d_x u$ is called the partial differential of u with respect to x; $d_y u$ the partial differential of u with respect to y; du the total differential of u. The equation $du = d_x u + d_y u$ expresses the fact that the total differential equals the sum of the partial differentials.

For example, if

$$u = x^2 + y^2 - xy, \quad d_x u = \frac{\partial u}{\partial x} dx = (2x - y) dx;$$

$$d_y u = \frac{\partial u}{\partial y} dy = (2y - x) dy;$$

$$du = \frac{\partial u}{\partial x} dx + \frac{\partial u}{\partial y} dy = (2x - y) dx + (2y - x) dy.$$

The differential expression (4) $du = Pdx + Qdy$, where P and Q are functions of the independent variables x and y, is said to be exact if it can be obtained by differentiating some function (5) $u = f(x, y)$. The differential of (5) is (6) $du = \frac{\partial u}{\partial x} dx + \frac{\partial u}{\partial y} dy$, and if (6) is identical with (4), (7) $\frac{\partial u}{\partial x} = P$, (8) $\frac{\partial u}{\partial y} = Q$. The partial y-derivative of (7) is $\frac{\partial^2 u}{\partial y \partial x} = \frac{\partial P}{\partial y}$; the partial x-derivative of (8) is $\frac{\partial^2 u}{\partial x \partial y} = \frac{\partial Q}{\partial x}$. Since $\frac{\partial^2 u}{\partial y \partial x} = \frac{\partial^2 u}{\partial x \partial y}$, the hypothesis that (4) is exact leads to the condition $\frac{\partial P}{\partial y} = \frac{\partial Q}{\partial x}$.

Conversely, if $\frac{\partial P}{\partial y} = \frac{\partial Q}{\partial x}$, the differential expression $du = Pdx + Qdy$ can be integrated. All the terms of u which contain x are found by integrating Pdx, considering y constant. That is, (1) $u = \int Pdx + Y$, where Y stands for the terms of u which do not contain x. To find Y from the y-derivative of (1), $\dfrac{\partial u}{\partial y} = \dfrac{\partial}{\partial y}\int Pdx + \dfrac{dY}{dy}$. If the expression $du = Pdx + Qdy$ can be integrated, $\dfrac{\partial u}{\partial y} = Q$. Hence $\dfrac{\partial}{\partial y}\int Pdx + \dfrac{dY}{dy} = Q$, and $\dfrac{dY}{dy} = Q - \dfrac{\partial}{\partial y}\int Pdx$. Y can be found by integration if $Q - \dfrac{\partial}{\partial y}\int Pdx$ is independent of x. Forming the x-derivative of this expression,

$$\frac{\partial}{\partial x}\left(Q - \frac{\partial}{\partial y}\int Pdx\right) = \frac{\partial Q}{\partial x} - \frac{\partial}{\partial x}\frac{\partial}{\partial y}\int Pdx$$

$$= \frac{\partial Q}{\partial x} - \frac{\partial}{\partial y}\frac{\partial}{\partial x}\int Pdx = \frac{\partial Q}{\partial x} - \frac{\partial P}{\partial y} = 0,$$

by hypothesis. Hence Y can be found by integration and the integral of the given differential expression becomes known.

Consider, for example, the differential expression

$$du = (3xy^2 - x^2)\,dx - (1 + 6y^2 - 3x^2y)\,dy.$$

This can be integrated since $\dfrac{\partial P}{\partial y} = 6xy = \dfrac{\partial Q}{\partial x}$. Integrating the first term of du, considering y constant, $u = \tfrac{3}{2}x^2y^2 - \tfrac{1}{3}x^3 + Y$. Differentiating partially with respect to y,

$$\frac{\partial u}{\partial y} = 3x^2y + \frac{dY}{dy} = -1 - 6y^2 + 3x^2y.$$

Whence $\quad \dfrac{dY}{dy} = -1 - 6y^2 \quad$ and $\quad Y = -y - 2y^3 + C.$

Finally $\quad u = \tfrac{3}{2}x^2y^2 - \tfrac{1}{3}x^3 - y - 2y^3 + C.$

PROBLEMS

Form the total differential of

1. $u = xy$. 2. $u = \dfrac{x}{y}$. 3. $u = x^3y$. 4. $u = \dfrac{x^2}{y}$.

5. The pressure of a gas on the containing vessel varies directly as the temperature and inversely as the volume; that is, $p = c\dfrac{t}{v}$, where c is a constant. Find the change of pressure when the temperature remains constant; the change of pressure when the volume remains constant; the change of pressure when temperature and volume both change.

Integrate the differential expressions:

6. $du = (3\,x^2 + 2\,ax)\,dx + (ax^2 + 3\,y^2)\,dy$.

7. $du = (x^3 + 3\,xy^2)\,dx + (y^3 + 3\,x^2y)\,dy$.

8. $du = (x^2 - 4\,xy - 2\,y^2)\,dx + (y^2 - 4\,xy - 2\,x^2)\,dy$.

9. $u = ax^2y^3$. Show that $x\dfrac{\partial u}{\partial x} + y\dfrac{\partial u}{\partial y} = 5\,u$.

10. $u = ax^2y^3 + bxy^4$. Show that $x\dfrac{\partial u}{\partial x} + y\dfrac{\partial u}{\partial y} = 5\,u$.

11. $u = F(x, y)$, where $F(x, y)$ is homogeneous of degree n. Show that $x\dfrac{\partial u}{\partial x} + y\dfrac{\partial u}{\partial y} = n \cdot u$.

The terms of $F(x, y)$ are of the general form $u_r = Ax^{(n-r)}y^r$. For all terms of this form $x\dfrac{\partial u_r}{\partial x} + y\dfrac{\partial u_r}{\partial y} = n \cdot u_r$, and the truth of the proposition follows directly. This is Euler's theorem on homogeneous functions.

12. If $du = f_1(x, y)\,dx + f_2(x, y)\,dy$ is exact and homogeneous of order $n - 1$, show that $n \cdot u = xf_1(x, y) + yf_2(x, y) + C$.

Denoting the integral of the given differential expression by $u = f(x, y)$, where $f(x, y)$ must be homogeneous of degree

n, since differentiation diminishes by unity the degree of an expression, $du = \frac{\partial u}{\partial x} dx + \frac{\partial u}{\partial y} dy$ and $n \cdot u = x \frac{\partial u}{\partial x} + y \frac{\partial u}{\partial y}$. In the given expression $\frac{\partial u}{\partial x} = f_1(x, y)$, $\frac{\partial u}{\partial y} = f_2(x, y)$. Hence

$$n \cdot u = x f_1(x, y) + y f_2(x, y).$$

Integrate the following expressions:

13. $du = (2 y^2 x + 3 y^3) dx + (2 x^2 y + 9 xy^2 + 8 y^3) dy.$
14. $du = (y^2 + 6 xy) dx + (2 xy + 3 x^2) dy.$
15. $du = (\tfrac{1}{2} x^{-\tfrac{1}{2}} y + 6 y^{\tfrac{1}{2}}) dx + (x^{\tfrac{1}{2}} + 3 xy^{-\tfrac{1}{2}}) dy.$

ART. 34. — DIFFERENTIATION OF INDIRECT FUNCTIONS

If $z = f(x, y)$ and $y = F(x)$, z is said to be a function of x directly and also indirectly through y. Denoting by Δx, Δy, and Δz the corresponding changes in the values of x, y, and z, $\Delta z = f(x + \Delta x, y + \Delta y) - f(x, y)$ and $\Delta y = F(x + \Delta x) - F(x)$, whence

$$\frac{\Delta z}{\Delta x} = \frac{f(x + \Delta x, y + \Delta y) - f(x, y + \Delta y)}{\Delta x} + \frac{f(x, y + \Delta y) - f(x, y)}{\Delta y} \frac{\Delta y}{\Delta x}$$

and $\frac{\Delta y}{\Delta x} = \frac{F(x + \Delta x) - F(x)}{\Delta x}$. Passing to the limit when limit $\Delta x = 0$, $\frac{dz}{dx} = \frac{\partial z}{\partial x} + \frac{\partial z}{\partial y} \frac{dy}{dx}$ and $\frac{dy}{dx} = F'(x)$.

EXAMPLE. — A point moves along the intersection of the paraboloid $z = 3 x^2 + 5 y^2$ and the plane $y = 2 x$. Find the rate of change of z and x. Here $\frac{\partial z}{\partial x} = 6 x$, $\frac{\partial z}{\partial y} = 10 y$, and $\frac{dy}{dx} = 2$. Hence $\frac{dz}{dx} = 6 x + 20 y.$

PROBLEMS

Determine $\dfrac{dz}{dx}$ when

1. $x^2 + y^2 + z^2 = 25$ and $y = 2x + 3$.
2. $x^2 + y^2 + z^2 = 0$ and $2x - 3y = 0$.
3. $\dfrac{x^2}{a^2} + \dfrac{y^2}{b^2} + \dfrac{z^2}{c^2} = 1$ and $x^2 + y^2 = r^2$.
4. $\dfrac{x^2}{a^2} + \dfrac{y^2}{b^2} + \dfrac{z^2}{c^2} = 1$ and $y^3 - 3xy + x^3 = 0$.
5. Determine $\dfrac{\partial u}{\partial x}$ and $\dfrac{\partial u}{\partial y}$ when $u = z^3$ and $z = x + y$.

Art. 35. — Envelopes.

The equation $f(x, y, a) = 0$ represents a curve whatever the value of a. By assigning to a all real values an infinite system of curves is obtained. The locus to which every curve of this system is tangent is called the envelope of the system of curves. The equation of the envelope is to be determined.

Let (1) $f(x, y, a) = 0$ and (2) $f(x, y, a + \Delta a) = 0$ represent any two curves of the system. Their points of intersection approach the points of tangency of $f(x, y, a) = 0$ with the envelope when Δa approaches zero. Hence the envelope may also be defined as the locus of the ultimate intersections of

$$f(x, y, a) = 0$$
and $f(x, y, a + \Delta a) = 0$

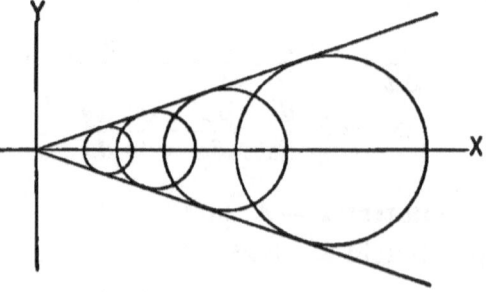

Fig. 34.

when Δa approaches zero. The points of intersection of (1) and

(2) satisfy $f(x, y, a) = 0$ and $\dfrac{f(x, y, a + \Delta a) - f(x, y, a)}{\Delta a} = 0$.
These equations, when Δa approaches zero, become

$$f(x, y, a) = 0 \text{ and } \frac{\partial}{\partial a} f(x, y, a) = 0.$$

The elimination of a from the last pair of equations gives the equation of the envelope.

In like manner the envelope of the singly infinite system of surfaces $f(x, y, z, a) = 0$ is found by eliminating a from $f(x, y, z, a) = 0$ and $\dfrac{\partial}{\partial a} f(x, y, z, a) = 0$.

The envelope of the doubly infinite system of surfaces $f(x, y, z, a, b) = 0$ is found by eliminating a and b from $f(x, y, z, a, b) = 0$, $\dfrac{\partial}{\partial a} f(x, y, z, a, b) = 0$, $\dfrac{\partial}{\partial b} f(x, y, z, a, b) = 0$.

EXAMPLE. — Find the envelope of the system of circles with centers on the X-axis and radii one-third of the distance of center from the origin.

The equation of the system of circles is $(x - a)^2 + y^2 = \dfrac{a^2}{9}$, where a is the distance of center from origin. Differentiating with respect to a, $-2(x - a) = \dfrac{2a}{9}$. Eliminating a from the equations $(x - a)^2 + y^2 = \dfrac{a^2}{9}$ and $-2(x - a) = \dfrac{2a}{9}$, $y^2 = \tfrac{1}{8} x^2$, the equation of the envelope. This equation represents the two straight lines $y = \dfrac{1}{2\sqrt{2}} x$ and $y = -\dfrac{1}{2\sqrt{2}} x$.

PROBLEMS

1. Find the envelope of the system of lines $\dfrac{x}{a} + \dfrac{y}{b} = 1$ when $a + b = c$, where c is a fixed constant.

2. Show that the envelope of the normals to the parabola $y^2 = 2px$ is a semi-cubic parabola.

3. Find the envelope of the system of ellipses $\dfrac{x^2}{a^2} + \dfrac{y^2}{b^2} = 1$ for which the area πab is constant.

4. Find the envelope of the system of ellipses $\dfrac{x^2}{a^2} + \dfrac{y^2}{b^2} = 1$ for which $a + b = c$, where c is a fixed constant.

5. Find the envelope of the system of planes $\dfrac{x}{a} + \dfrac{y}{b} + \dfrac{z}{c} = 1$ when $abc = m$, where m is a fixed constant.

6. Find the envelope of the system of spheres whose centers lie in the XY-plane and whose radii vary as the distance from origin to center.

CHAPTER VII

CIRCULAR AND INVERSE CIRCULAR FUNCTIONS

ART. 36. — DIFFERENTIATION OF CIRCULAR FUNCTIONS

Denote by x the circular measure of an angle less than a right angle. From the figure triangle OPD < sector OPA < triangle OTA. Hence

$$\tfrac{1}{2}r^2 \cdot \sin x \cdot \cos x < \tfrac{1}{2}r^2 \cdot x < \tfrac{1}{2}r^2 \cdot \tan x$$

and $\cos x < \dfrac{x}{\sin x} < \dfrac{1}{\cos x}$. This inequality is true for all values of x less than $\dfrac{\pi}{2}$. When x approaches zero, $\cos x$ approaches unity. Since $\dfrac{x}{\sin x}$ lies between two numbers, $\cos x$ and $\dfrac{1}{\cos x}$, whose common limit is unity when x approaches zero, limit $\dfrac{x}{\sin x} = 1$ when limit $x = 0$. This limit is fundamental in this chapter.

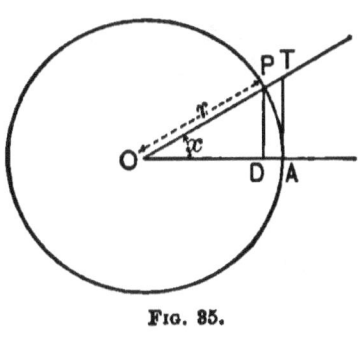

FIG. 85.

Denote by u a continuous function of x which changes by Δu when x changes by Δx. Then

I. $\dfrac{d}{dx}\sin u = \text{limit} \dfrac{\sin(u + \Delta u) - \sin u}{\Delta u} \cdot \dfrac{\Delta u}{\Delta x}$

$= \text{limit} \dfrac{2\sin\tfrac{1}{2}\Delta u \cdot \cos\tfrac{1}{2}(2u + \Delta u)}{\Delta u} \cdot \dfrac{\Delta u}{\Delta x}$

CIRCULAR FUNCTIONS

$$= \text{limit}\, \frac{\sin\frac{\Delta u}{2}}{\frac{\Delta u}{2}} \cdot \cos\left(u + \frac{\Delta u}{2}\right) \cdot \frac{\Delta u}{\Delta x}$$

$$= \cos u \cdot \frac{du}{dx} \text{ when limit } \Delta x = 0.$$

II. $\dfrac{d}{dx}\cos u = \dfrac{d}{dx}\sin\left(\dfrac{\pi}{2} - u\right) = \cos\left(\dfrac{\pi}{2} - u\right)\cdot\dfrac{d}{dx}\left(\dfrac{\pi}{2} - u\right)$

$$= -\sin u \cdot \frac{du}{dx}.$$

III. $\dfrac{d}{dx}\tan u = \dfrac{d}{dx}\dfrac{\sin u}{\cos u} = \dfrac{\cos u \cdot \dfrac{d}{dx}\sin u - \sin u \cdot \dfrac{d}{dx}\cos u}{\cos^2 u}$

$$= \frac{\cos^2 u \cdot \frac{du}{dx} + \sin^2 u \cdot \frac{du}{dx}}{\cos^2 u} = \frac{\frac{du}{dx}}{\cos^2 u} = \sec^2 u \cdot \frac{du}{dx}.$$

IV. $\dfrac{d}{dx}\cot u = \dfrac{d}{dx}\tan\left(\dfrac{\pi}{2} - u\right) = \sec^2\left(\dfrac{\pi}{2} - u\right)\cdot\dfrac{d}{dx}\left(\dfrac{\pi}{2} - u\right)$

$$= -\operatorname{cosec}^2 u \cdot \frac{du}{dx}.$$

V. $\dfrac{d}{dx}\sec u = \dfrac{d}{dx}\dfrac{1}{\cos u} = \dfrac{\sin u \cdot \dfrac{du}{dx}}{\cos^2 u}$

$$= \frac{\sin u}{\cos u}\frac{1}{\cos u}\frac{du}{dx} = \tan u \cdot \sec u \cdot \frac{du}{dx}.$$

VI. $\dfrac{d}{dx}\operatorname{cosec} u = \dfrac{d}{dx}\sec\left(\dfrac{\pi}{2} - u\right)$

$$= \tan\left(\frac{\pi}{2} - u\right)\cdot\sec\left(\frac{\pi}{2} - u\right)\frac{d}{dx}\left(\frac{\pi}{2} - u\right)$$

$$= -\cot u \cdot \operatorname{cosec} u \cdot \frac{du}{dx}.$$

VII. $\dfrac{d}{dx}$ vers $u = \dfrac{d}{dx}(1 - \cos u) = \sin u \cdot \dfrac{du}{dx}$.

VIII. $\dfrac{d}{dx}$ covers $u = \dfrac{d}{dx}(1 - \sin u) = -\cos u \cdot \dfrac{du}{dx}$.

EXAMPLE I. — Differentiate $\cos(3x^2)$.

$\dfrac{d}{dx}\cos(3x^2) = -\sin(3x^2) \cdot \dfrac{d}{dx}(3x^2) = -6x \cdot \sin(3x^2)$.

EXAMPLE II. — Assuming that the relative rate of change of $\cos x$ and x remains the same as at 30° throughout the next 10', calculate cos 30° 10',

$\sin 30° = .5$, $\cos 30° = .86603$, and $\dfrac{d}{dx}\cos x = -\sin x$.

Hence at 30° the relative rate of change of $\cos x$ and x is $-.5$. Since x is the circular measure of the angle, it is necessary to express the increment of the angle in circular measure. The circular measure of an angle of 10' is $\dfrac{3.14159}{180 \times 6} = .0029088$. To a change of .0029088 in the value of x there corresponds a change of $-.0014544$ in the value of $\cos x$ under the assumption of the problem. Hence $\cos 30° 10' = .86458$.

PROBLEMS

Differentiate,

1. $\sin(x^2)$.
2. $\sin(5x)$.
3. $\sin(7x)$.
4. $2(\sin x)^2$.
5. $\sin(3x)^2$.
6. $4\sin\left(\dfrac{1}{x}\right)$.
7. $5\cos^2(2x)$.
8. $\sin x \cdot \cos(3x)$.
9. $7x \cdot \tan(3x)$.
10. $\cot^2(2x)$.
11. $\sec(x^2)$.
12. $\operatorname{covers}(3x+5)$.
13. $\sin(2x-7)$.
14. $\tan(3-4x)$.
15. $\cos^{\frac{1}{2}}(5x)$.
16. $\tan(x^{\frac{1}{2}})$.
17. $\sin^7 x$.
18. $\cos^4 x$.
19. $\sin(3x)^{\frac{1}{2}}$.
20. $\tan^3 x$.
21. $\cot^5 x$.

CIRCULAR FUNCTIONS

22. Find the rate of change of $\sin x$ when $x = 45°$.

23. Find the rate of change of $\tan x$ when $x = 45°$.

24. Find the rate of change of $\cos x$ when $x = 240°$.

25. Find when $\sin x$ changes half as fast as x.

26. Find when $\cos x$ decreases half as fast as x increases.

27. Find when $\tan x$ changes four times as fast as x.

28. Assuming the rate of change of $\sin x$ to remain the same as at 60° throughout the next 5', calculate $\sin 60° 5'$.

29. Assuming the rate of change of $\tan x$ to remain the same as at 45° throughout the next 2', calculate $\tan 45° 2'$.

30. $\dfrac{dy}{dx} = (a^2 - x^2)^{\frac{1}{2}}$ and $x = a \cdot \sin \theta$. Find $\dfrac{dy}{d\theta}$.

Since $\dfrac{dy}{d\theta} = \dfrac{dy}{dx} \cdot \dfrac{dx}{d\theta}$ and $\dfrac{dy}{dx} = (a^2 - x^2)^{\frac{1}{2}} = a \cos \theta$, $\dfrac{dx}{d\theta} = a \cos \theta$,

$\dfrac{dy}{d\theta} = a^2 \cdot \cos^2 \theta$.

31. $\dfrac{dy}{dx} = x(x^2 + a^2)^{\frac{1}{2}}$. Find $\dfrac{dy}{d\theta}$ when $x = a \cdot \tan \theta$.

32. $\dfrac{dy}{dx} = x^2(x^2 - a^2)^{\frac{1}{2}}$. Find $\dfrac{dy}{d\theta}$ when $x = a \cdot \sec \theta$.

33. $\dfrac{dy}{dx} = x^2(2ax - x^2)^{\frac{1}{2}}$. Find $\dfrac{dy}{d\theta}$ when $x = a(1 - \cos \theta)$.

34. Form $\dfrac{dy}{dx}$ and $\dfrac{d^2y}{dx^2}$ for circle $x = R \cdot \cos \theta$, $y = R \cdot \sin \theta$.

Since x and y are continuous functions of θ, if Δx, Δy, and $\Delta \theta$ denote corresponding changes in x, y, and θ, limit $\Delta x = 0$ and limit $\Delta y = 0$ when limit $\Delta \theta = 0$. For all values of $\Delta \theta$,

$\dfrac{\Delta y}{\Delta x} = \dfrac{\frac{\Delta y}{\Delta \theta}}{\frac{\Delta x}{\Delta \theta}}$. Hence when limit $\Delta \theta = 0$, $\dfrac{dy}{dx} = \dfrac{\frac{dy}{d\theta}}{\frac{dx}{d\theta}}$.

35. Form $\dfrac{dy}{dx}$ and $\dfrac{d^2y}{dx^2}$ for cycloid $x = R\theta - R\sin\theta$, $y = R - R\cos\theta$.

36. Form $\dfrac{dy}{dx}$ and $\dfrac{d^2y}{dx^2}$ for cycloid $x = R - R\cos\theta$, $y = R\theta + R\sin\theta$.

37. Show that $2\tan x - \tan^2 x$ has a maximum value when $x = \dfrac{\pi}{4}$.

38. Show that $\tan x + 3\cot x$ has a minimum value when $x = \dfrac{\pi}{3}$.

39. Find the maximum radii vectores of $r = a\sin(3\theta)$.

40. Examine $\sin x \cos^3 x$ for maxima and minima.

41. Examine $\sin x + \cos x$ for maxima and minima.

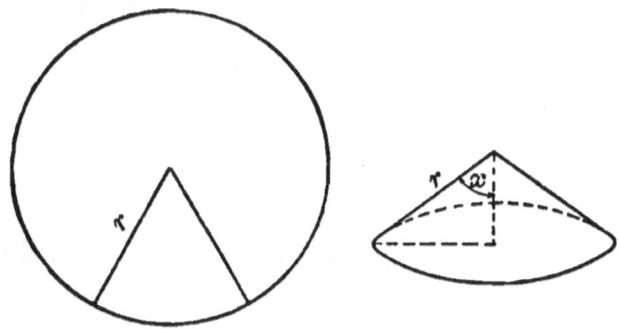

Fig. 86.

42. Find the length of the arc of the sector which must be cut from a circular piece of sheet iron so that the remainder may form a conical vessel of maximum capacity.

$$V = \frac{\pi}{3} r^3 \sin^2 x \cos x.$$

43. A steamer whose speed is 8 knots per hour and course due north sights another steamer directly ahead whose speed is 10 knots and course due west. What course must the first steamer take to cross the track of the second steamer at the least possible distance from her?

Find the true values of,

44. $\dfrac{1-\cos x}{x^2}$ when $x=0$.   47. $\dfrac{x-\sin x\cos x}{x^3}$ when $x=0$.

45. $\dfrac{x-\sin x}{x^3}$ when $x=0$.   48. $\dfrac{\tan x - x}{x-\sin x}$ when $x=0$.

46. $\dfrac{\sin x}{x}$ when $x=0$.   49. $\dfrac{x\sin x}{x-2\sin x}$ when $x=0$.

50. Show that the radius of curvature of the cycloid $x = R\theta - R\sin\theta$, $y = R - R\cos\theta$ is twice the normal.

ART. 37.—EVALUATION OF THE FORMS $\infty \cdot 0$, $\infty - \infty$, $\dfrac{\infty}{\infty}$

If $f(x)\cdot \phi(x)$ when $x=a$ takes the form $\infty \cdot 0$,

$$f(x)\cdot\phi(x) = \dfrac{\phi(x)}{\dfrac{1}{f(x)}} = \dfrac{0}{0} \text{ when } x=a,$$

and the true value of the expression may be found by the method of Art. 26.

If $f(x) - \phi(x)$ takes the form $\infty - \infty$ when $x=a$,

$$f(x) - \phi(x) = \dfrac{1}{\dfrac{1}{f(x)}} - \dfrac{1}{\dfrac{1}{\phi(x)}} = \dfrac{\dfrac{1}{\phi(x)} - \dfrac{1}{f(x)}}{\dfrac{1}{f(x)}\dfrac{1}{\phi(x)}} = \dfrac{0}{0} \text{ when } x=a,$$

and again the method of Art. 26 may be applied. The reduction of the form $\infty - \infty$ to the form $\dfrac{0}{0}$ can frequently be effected more directly.

If $y = \dfrac{f(x)}{\phi(x)} = \dfrac{\infty}{\infty}$ when $x = a$, $y = \dfrac{\dfrac{1}{\phi(x)}}{\dfrac{1}{f(x)}} = \dfrac{0}{0}$ when $x = a$, and the method of Art. 26 gives

$$y = \dfrac{\dfrac{-\phi'(x)}{\{\phi(x)\}^2}}{\dfrac{-f'(x)}{\{f(x)\}^2}} = \left\{\dfrac{f(x)}{\phi(x)}\right\}^2 \dfrac{\phi'(x)}{f'(x)} \text{ when } x = a.$$

Hence, $y = y^2 \dfrac{\phi'(x)}{f'(x)}$ and $y = \dfrac{f(x)}{\phi(x)} = \dfrac{f'(x)}{\phi'(x)}$ when $x = a$. That is, the form $\dfrac{\infty}{\infty}$ is evaluated in precisely the same manner as the form $\dfrac{0}{0}$.

EXAMPLE I. — Evaluate $(a^2 - x^2) \tan\dfrac{\pi x}{2a}$ when $x = a$.

$$(a^2 - x^2) \tan\dfrac{\pi x}{2a} = 0 \cdot \infty \text{ when } x = a.$$

$$(a^2 - x^2) \tan\dfrac{\pi x}{2a} = \dfrac{a^2 - x^2}{\cot\dfrac{\pi x}{2a}} = \dfrac{0}{0} \text{ when } x = a.$$

Hence, $(a^2 - x^2) \tan\dfrac{\pi x}{2a} = \dfrac{-2x}{-\dfrac{\pi}{2a}\cosec^2\dfrac{\pi x}{2a}} = \dfrac{4a^2}{\pi}$ when $x = a$.

EXAMPLE II. — Evaluate $\sec x - \tan x$ when $x = \dfrac{\pi}{2}$.

$$\sec x - \tan x = \infty - \infty \text{ when } x = \dfrac{\pi}{2}.$$

$$\sec x - \tan x = \dfrac{1 - \sin x}{\cos x} = \dfrac{0}{0} \text{ when } x = \dfrac{\pi}{2}.$$

Hence, $\sec x - \tan x = \dfrac{-\cos x}{-\sin x} = 0$ when $x = \dfrac{\pi}{2}$.

PROBLEMS

Evaluate, 1. $(1-x)\tan\dfrac{\pi x}{2}$ when $x=1$.

2. $\dfrac{\frac{1}{4}\dfrac{\pi}{x}}{\cot\dfrac{\pi x}{2}}$ when $x=0$.   3. $x\sin^2\dfrac{a}{x}$ when $x=\infty$.

4. $x\cot x$ when $x=0$.   5. $\sec(3x)\cos(5x)$ when $x=\dfrac{\pi}{2}$.

6. $2x\tan x - \pi\sec x$ when $x=\dfrac{\pi}{2}$.

7. $(1-\tan x)\sec(2x)$ when $x=\dfrac{\pi}{4}$.

ART. 38. — INTEGRATION OF CIRCULAR FUNCTIONS

The eight formulas of Art. 36 may be written,

I. $\displaystyle\int\cos u\cdot\dfrac{du}{dx}=\sin u + C,$ III. $\displaystyle\int\sec^2 u\cdot\dfrac{du}{dx}=\tan u + C,$

II. $\displaystyle\int\sin u\cdot\dfrac{du}{dx}=-\cos u + C,$ IV. $\displaystyle\int\operatorname{cosec}^2 u\cdot\dfrac{du}{dx}=-\cot u + C,$

V. $\displaystyle\int\tan u\cdot\sec u\cdot\dfrac{du}{dx}=\sec u + C,$

VI. $\displaystyle\int\cot u\cdot\operatorname{cosec} u\cdot\dfrac{du}{dx}=-\operatorname{cosec} u + C,$

VII. $\displaystyle\int\sin u\cdot\dfrac{du}{dx}=\operatorname{vers} u + C,$

VIII. $\displaystyle\int\cos u\cdot\dfrac{du}{dx}=-\operatorname{covers} u + C.$

EXAMPLE I. — Integrate $\dfrac{dy}{dx}=x\cdot\sin(2x^2)$.

This derivative has the general form $\sin u\cdot\dfrac{du}{dx}$. Placing $u=2x^2$, $\dfrac{du}{dx}=4x$, and the given derivative may be written

$\frac{dy}{dx} = \frac{1}{4} \cdot \sin(2x^2) \cdot 4x$. Hence, by the formula,

$$\int \sin u \cdot \frac{du}{dx} = -\cos u + C, \quad y = -\tfrac{1}{4}\cos(2x^2) + C.$$

EXAMPLE II. — Integrate $\frac{dy}{dx} = \sin^5 x \cdot \cos x$.

Since $\frac{d}{dx}\sin x = \cos x$, $\frac{dy}{dx} = \sin^5 x \cdot \frac{d}{dx}\sin x$, and $y = \tfrac{1}{6}\sin^6 x + C$.

PROBLEMS

Integrate,

1. $\frac{dy}{dx} = \cos(4x)$.

2. $\frac{dy}{dx} = \cos(2x - 5)$.

3. $\frac{dy}{dx} = x^2 \cdot \cos(x^3 + 2)$.

4. $\frac{dy}{dx} = \sec^2(3x + 5)$.

5. $\frac{dy}{dx} = \sec^2\left(\frac{x}{2}\right)$.

6. $\frac{dy}{dx} = \cos(\tfrac{3}{2}x) \cdot \cot(\tfrac{3}{2}x)$.

7. $\frac{dy}{dx} = \frac{1}{x^2} \cdot \cos\frac{1}{x}$.

8. $\frac{dy}{dx} = 3\sin(5x - 7)$.

9. $\frac{dy}{dx} = \sin^2 x \cdot \cos x$.

10. $\frac{dy}{dx} = \sin^{\frac{3}{2}} x \cdot \cos x$.

11. $\frac{dy}{dx} = \cos^3 x \cdot \sin x$.

12. $\frac{dy}{dx} = \tan^5 x \cdot \sec^2 x$.

Frequently an expression containing circular functions, when not directly integrable, may be transformed into an expression which can be integrated by means of the trigonometric relations $\sin^2 x + \cos^2 x = 1$, $\sec^2 x = 1 + \tan^2 x$, $\sin^2 x = \tfrac{1}{2} - \tfrac{1}{2}\cos(2x)$, $\cos^2 x = \tfrac{1}{2} + \tfrac{1}{2}\cos(2x)$, $\sin x \cos x = \tfrac{1}{2}\sin(2x)$.

The last three relations are special cases of the formulas

$$\sin x \sin y = \tfrac{1}{2}\cos(x - y) - \tfrac{1}{2}\cos(x + y),$$
$$\cos x \cos y = \tfrac{1}{2}\cos(x - y) + \tfrac{1}{2}\cos(x + y),$$
$$\sin x \cos y = \tfrac{1}{2}\sin(x + y) + \tfrac{1}{2}\sin(x - y),$$

which are frequently useful.

CIRCULAR FUNCTIONS

EXAMPLE I. — Integrate $\dfrac{dy}{dx} = \sin^5 x \cdot \cos^2 x$.

Write $\quad \dfrac{dy}{dx} = \sin^5 x \cdot \cos^2 x = \sin^4 x \cdot \cos^2 x \cdot \sin x$

$$= -(1 - \cos^2 x)^2 \cdot \cos^2 x \cdot \dfrac{d}{dx} \cos x$$

$$= -\cos^2 x \dfrac{d}{dx} \cos x + 2 \cos^4 x \cdot \dfrac{d}{dx} \cos x - \cos^6 x \cdot \dfrac{d}{dx} \cos x.$$

Integrating term by term, $y = -\tfrac{1}{3} \cos^3 x + \tfrac{2}{5} \cos^5 x - \tfrac{1}{7} \cos^7 x + C$.

EXAMPLE II. — Integrate $\dfrac{dy}{dx} = \tan^4 x$.

Write $\dfrac{dy}{dx} = \tan^4 x = \tan^2 x (\sec^2 x - 1)$

$$= \tan^2 x \cdot \sec^2 x - \tan^2 x = \tan^2 x \cdot \sec^2 x - \sec^2 x + 1.$$

Integrating term by term, $y = \tfrac{1}{3} \tan^3 x - \tan x + x + C$.

EXAMPLE III. — Integrate $\dfrac{dy}{dx} = \sin^4 x$.

Write $\dfrac{dy}{dx} = \sin^4 x = \{\tfrac{1}{2} - \tfrac{1}{2} \cos (2x)\}^2$

$$= \tfrac{1}{4} - \tfrac{1}{2} \cos (2x) + \tfrac{1}{4} \cos^2 (2x)$$

$$= \tfrac{1}{4} - \tfrac{1}{2} \cos (2x) + \tfrac{1}{4} \{\tfrac{1}{2} + \tfrac{1}{2} \cos (4x)\}$$

$$= \tfrac{3}{8} - \tfrac{1}{2} \cos (2x) + \tfrac{1}{8} \cos (4x).$$

Integrating term by term, $y = \tfrac{3}{8} x - \tfrac{1}{4} \sin (2x) + \tfrac{1}{32} \cos (4x) + C$.

EXAMPLE IV. — Integrate $\dfrac{dy}{dx} = \dfrac{\sin^2 x}{\cos^4 x}$.

Write $\dfrac{dy}{dx} = \dfrac{\sin^2 x}{\cos^4 x} = \dfrac{\sin^2 x}{\cos^2 x} \cdot \dfrac{1}{\cos^2 x} = \tan^2 x \cdot \sec^2 x$. Integrating, $y = \tfrac{1}{3} \tan^3 x + C$.

EXAMPLE V. — Integrate $\frac{dy}{dx} = \sin^2 x \cdot \cos^2 x$.

Write $\frac{dy}{dx} = \sin^2 x \cdot \cos^2 x = \frac{1}{4} \sin^2 (2x)$

$\qquad = \frac{1}{4} \{\frac{1}{2} - \frac{1}{2} \cos (4x)\} = \frac{1}{8} - \frac{1}{8} \cos (4x)$.

Integrating, $y = \frac{1}{8} x - \frac{1}{32} \sin (4x) + C$.

EXAMPLE VI. — Integrate $\frac{dy}{dx} = \sin x \cos 4x$.

Write $\frac{dy}{dx} = \sin x \cos 4x = \frac{1}{2} \sin 5x - \frac{1}{2} \sin 3x$. Integrating, $y = -\frac{1}{10} \cos 5x + \frac{1}{6} \cos 3x + C$.

PROBLEMS

Integrate,

1. $\frac{dy}{dx} = \sin^2 (3x) \cdot \cos^3 (3x)$.
2. $\frac{dy}{dx} = \sin^3 (\frac{1}{2} x)$.
3. $\frac{dy}{dx} = \cos^2 (2x)$.
4. $\frac{dy}{dx} = \tan^3 (2x)$.
5. $\frac{dy}{dx} = 3 \cos^3 x$.
6. $\frac{dy}{dx} = 5 \sin^2 (\frac{1}{2} x)$.
7. $\frac{dy}{dx} = \sin^3 x \cdot \cos^2 x$.
8. $\frac{dy}{dx} = \sin^3 x \cdot \cos^3 x$.
9. $\frac{dy}{dx} = \sin^4 x \cdot \cos^4 x$.
10. $\frac{dy}{dx} = \tan^5 x$.
11. $\frac{dy}{dx} = \cos^5 (3x)$.
12. $\frac{dy}{dx} = 3x \cdot \sin^2 (x^2) \cdot \cos^3 (x^2)$.
13. $\frac{dy}{dx} = x^2 \cdot \sec^2 (3 x^3)$.
14. $\frac{dy}{dx} = \frac{\sin^5 x}{\cos^2 x}$.

Show that, m and n being positive integers,

15. $\int_0^{2\pi} \sin (mx) \cdot dx = 0$.
16. $\int_0^{2\pi} \cos (mx) \cdot dx = 0$.

CIRCULAR FUNCTIONS

17. $\int_0^{2\pi} \sin(mx) \cdot \sin(nx) \cdot dx = 0, \ m \neq n.$

18. $\int_0^{2\pi} \cos(mx) \cdot \cos(nx) \cdot dx = 0, \ m \neq n.$

19. $\int_0^{2\pi} \cos(mx) \cdot \sin(nx) \cdot dx = 0.$

20. $\int_0^{2\pi} \cos^2(mx) \cdot dx = \pi.$ 21. $\int_0^{2\pi} \sin^2(mx) \cdot dx = \pi.$

ART. 39. — INTEGRATION BY TRIGONOMETRIC SUBSTITUTION

First derivatives with respect to x involving $\sqrt{a^2 - x^2}$ may be transformed into trigonometric derivatives by the substitution $x = a \cdot \cos\theta$; those involving $\sqrt{a^2 + x^2}$ by the substitution $x = a \cdot \tan\theta$; those involving $\sqrt{x^2 - a^2}$ by the substitution $x = a \cdot \sec\theta$; those involving $\sqrt{2ax - x^2}$ by the substitution $x = a(1 - \cos\theta)$.

EXAMPLE. — Integrate $\dfrac{dy}{dx} = \dfrac{1}{x^3(x^2 - a^2)^{\frac{1}{2}}}.$

Placing $x = a \cdot \sec\theta$, $\dfrac{dx}{d\theta} = a \cdot \sec\theta \cdot \tan\theta$, $(x^2 - a^2)^{\frac{1}{2}} = a \cdot \tan\theta.$

Hence, $\dfrac{dy}{d\theta} = \dfrac{dy}{dx} \cdot \dfrac{dx}{d\theta} = \dfrac{1}{a^3} \cdot \cos^2\theta = \dfrac{1}{a^3}\{\tfrac{1}{2} + \tfrac{1}{2}\cos(2\theta)\}.$

Integrating, $y = \dfrac{1}{a^3}\{\tfrac{1}{2}\theta + \tfrac{1}{4}\sin(2\theta)\} + C.$ From $x = a \cdot \sec\theta$,

$\theta = \sec^{-1}\dfrac{x}{a}$, $\sin(2\theta) = 2\sin\theta\cos\theta = 2\dfrac{a}{x}\left(1 - \dfrac{a^2}{x^2}\right)^{\frac{1}{2}} = \dfrac{2a}{x^2}(x^2 - a^2)^{\frac{1}{2}}.$

Finally, $y = \dfrac{1}{2a^3} \cdot \sec^{-1}\dfrac{x}{a} + \dfrac{1}{2a^2x^2}(x^2 - a^2)^{\frac{1}{2}} + C.$

PROBLEMS

Integrate,

1. $\dfrac{dy}{dx} = \dfrac{1}{(a^2 - x^2)^{\frac{3}{2}}}.$ 2. $\dfrac{dy}{dx} = \dfrac{1}{x^2(1 + x^2)^{\frac{1}{2}}}.$

3. $\dfrac{dy}{dx} = \dfrac{1}{x^3(x^2-1)^{\frac{1}{2}}}.$

4. $\dfrac{dy}{dx} = \dfrac{x^3}{(1-x^2)^{\frac{1}{2}}}.$

5. $\dfrac{dy}{dx} = \dfrac{3}{(2+x^2)^{\frac{3}{2}}}.$

6. $\dfrac{dy}{dx} = \dfrac{x^5}{(1-x^2)^{\frac{1}{2}}}.$

7. $\dfrac{dy}{dx} = \dfrac{x^5}{\sqrt{1-x^2}}.$

8. $\dfrac{dy}{dx} = \dfrac{x^5}{\sqrt{1+x^2}}.$

9. $\dfrac{dy}{dx} = x^3(1-x^2)^{\frac{1}{2}}.$

10. $\dfrac{dy}{dx} = \dfrac{1}{(a^2+x^2)^2}.$ Substitute $x = a \cdot \tan\theta$.

11. $\dfrac{dy}{dx} = \dfrac{1}{(1-x^2)^2}.$

12. $\dfrac{dy}{dx} = \sqrt{\dfrac{1-x}{x}}.$ Substitute $x^{\frac{1}{2}} = \sin\theta$.

13. Find the area of the circle $x^2 + y^2 = r^2$.

$A = 2\displaystyle\int_{-r}^{+r} (r^2 - x^2)^{\frac{1}{2}} \cdot dx.$ Substituting $x = r \cdot \cos\theta$, and noticing that when x runs through all values from $+r$ to $-r$, θ runs from $+\dfrac{\pi}{2}$ to $-\dfrac{\pi}{2}$, there results

$$A = r^2 \int_{-\frac{\pi}{2}}^{+\frac{\pi}{2}} \cos^2\theta \cdot d\theta = \pi \cdot r^2.$$

14. Find the area of the ellipse $\dfrac{x^2}{a^2} + \dfrac{y^2}{b^2} = 1.$

15. Find the area of the hypocycloid $x^{\frac{2}{3}} + y^{\frac{2}{3}} = a^{\frac{2}{3}}$

16. Find the volume of the solid generated by the revolution of $y = \dfrac{8a^3}{x^2 + 4a^2}$ about the X-axis.

17. Find the area bounded by the curve $y^2 = \dfrac{x^3}{2a-x}$ and the line $x = 2a$. $A = \displaystyle\int_0^{2a} \dfrac{x^{\frac{3}{2}}\, dx}{(2a-x)^{\frac{1}{2}}}.$ Substitute $x = 2a \cdot \sin^2\theta$.

Art. 40. — Polar Curves

Tangents and Normals. — Let $r = f(\theta)$ be the equation of a continuous plane curve, $P(r, \theta)$ any point in the curve, and $P'(r + \Delta r, \theta + \Delta\theta)$ any other point in the curve. The ratio $\dfrac{\Delta r}{\Delta \theta}$ measures the average rate of change of r in the interval $\Delta\theta$, and $\dfrac{dr}{d\theta} = \text{limit} \dfrac{\Delta r}{\Delta \theta}$ when limit $\Delta\theta = 0$ measures the actual rate of change of r at the point (r, θ).

The secant through (r, θ), $(r + \Delta r, \theta + \Delta\theta)$ approaches the tangent to the curve at (r, θ) as $\Delta\theta$ approaches zero. Hence the angle $AP'S$ approaches the angle $APT = \phi$ included by the tangent at (r, θ), and the radius vector to the point (r, θ) when $\Delta\theta$ approaches zero. Drawing a perpendicular from (r, θ) to AP',

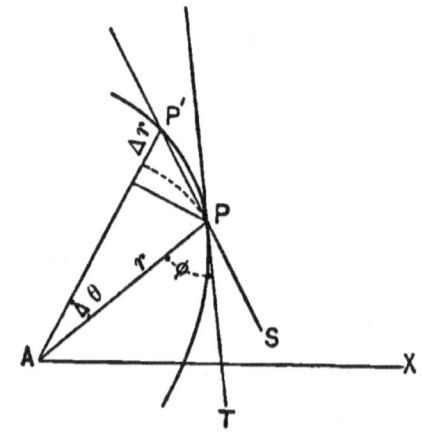

Fig. 37.

$$\tan\phi = \text{limit} \tan AP'S = \text{limit} \frac{r \cdot \sin\Delta\theta}{r + \Delta r - r \cdot \cos\Delta\theta}$$

$$= \text{limit} \frac{r \cdot \dfrac{\sin\Delta\theta}{\Delta\theta}}{\dfrac{r \sin\dfrac{\Delta\theta}{2}}{\dfrac{\Delta\theta}{2}} \sin\dfrac{\Delta\theta}{2} + \dfrac{\Delta r}{\Delta\theta}}$$

$$= r \cdot \frac{d\theta}{dr} \quad \text{when limit } \Delta\theta = 0.$$

The distance AT from the pole to the tangent measured on the perpendicular to the radius vector to the point of tangency

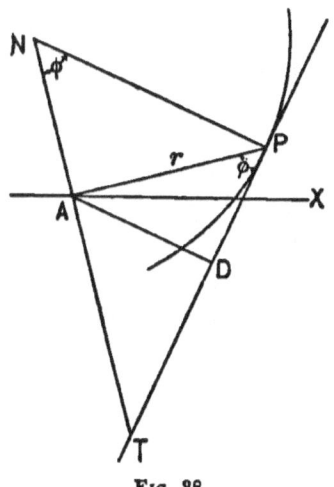

Fig. 88.

is called the polar subtangent; the distance AN from the pole to the normal measured on the same perpendicular is called the polar subnormal. From the figure subnormal $AN = \dfrac{dr}{d\theta}$; normal $PN = \left(r^2 + \dfrac{dr^2}{d\theta^2}\right)^{\frac{1}{2}}$; subtangent $AT = r^2 \cdot \dfrac{d\theta}{dr}$; perpendicular from pole to tangent

$$AD = p = \dfrac{r^2}{r^2 + \left(\dfrac{dr^2}{d\theta^2}\right)^{\frac{1}{2}}}.$$

Asymptotes. — When in a polar curve $r = f(\theta)$, for some finite value of $\theta = \theta_0$, r becomes infinite and the subtangent is finite, the tangent to the curve at infinity passes at a finite distance from the pole and is called an asymptote. If the subtangent is positive, lay it off to the right of the radius vector looking towards the infinitely distant point of the curve; if negative, lay off the subtangent to the left of this radius vector.

EXAMPLE. — Examine $r = \dfrac{a}{\theta}$ for asymptotes.

For $\theta = 0$, $r = \infty$. The subtangent $= r^2 \cdot \dfrac{d\theta}{dr} = -a$. Hence

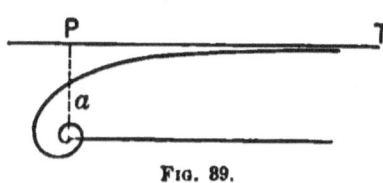

Fig. 89.

for $\theta = 0$, subtangent $= -a$, and the asymptote is obtained by laying off on the perpendicular to the polar axis at the origin to the left when facing in the direction $\theta = 0$ the distance a and drawing the perpendicular PT.

CIRCULAR FUNCTIONS

PROBLEMS

1. Find the subtangent of $r = a \cdot \theta$.
2. Find the subtangent of $r^2 = a^2 \cdot \cos(2\theta)$.
3. Examine $r = \dfrac{a}{\theta^{\frac{1}{2}}}$ for asymptotes.
4. Find the angle under which the radius vector cuts $r = \dfrac{p}{1 - \cos\theta}$.
5. Show that the radius vector cuts $r = a^\theta$ under a constant angle.

Length. — Denoting the length of the curve $r = f(\theta)$ from $\theta = \theta_0$ to $\theta = \theta_1$, by s

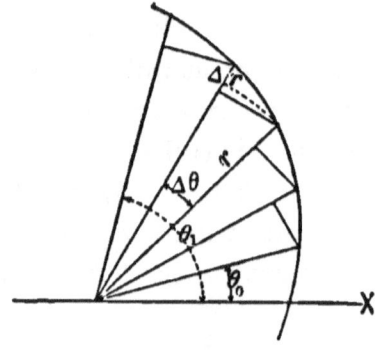

Fig. 40.

$$s = \operatorname*{limit}_{\theta=\theta_0}^{\theta=\theta_1} \Sigma \sqrt{r^2 \cdot \sin^2\Delta\theta + (r + \Delta r - r \cdot \cos\Delta\theta)^2}$$

$$= \operatorname*{limit}_{\theta=\theta_0}^{\theta=\theta_1} \Sigma \sqrt{r^2\left(\frac{\sin\Delta\theta}{\Delta\theta}\right)^2 + \left\{\frac{r(1 - \cos\Delta\theta) + \Delta r}{\Delta\theta}\right\}^2} \Delta\theta$$

$$= \int_{\theta=\theta_0}^{\theta=\theta_1} \sqrt{r^2 + \frac{dr^2}{d\theta^2}} \cdot d\theta, \text{ when limit } \Delta\theta = 0.$$

Area. — Denoting the area bounded by the curve $r = f(\theta)$ and the radii vectores to the points (r_0, θ_0), (r_1, θ_1) by A,

$$A = \text{limit} \sum_{\theta=\theta_0}^{\theta=\theta_1} \tfrac{1}{2} r^2 \cdot \Delta\theta = \tfrac{1}{2} \int_{\theta_0}^{\theta_1} r^2 \cdot d\theta.$$

EXAMPLE. — Find the length and the area of the cardioid $r = a(1 + \cos\theta)$.

The length

$$s = 2\int_0^\pi \left(r^2 + \frac{dr^2}{d\theta^2}\right)^{\frac{1}{2}} \cdot d\theta = 2\int_0^\pi \left\{(a + a\cdot\cos\theta)^2 + a^2\cdot\sin^2\theta\right\}^{\frac{1}{2}} d\theta$$

$$= 2a\int_0^\pi \sqrt{2(1+\cos\theta)}\cdot d\theta = 4a\cdot\cos\tfrac{1}{2}\theta\cdot d\theta = 8a(\sin\tfrac{1}{2}\theta)_0^\pi = 8a.$$

The area $A = 2\cdot\tfrac{1}{2}\int_0^\pi r^2\cdot d\theta = a^2\int_0^\pi (1+\cos\theta)^2\cdot d\theta$

$$= a^2\int_0^\pi (1 + 2\cos\theta + \cos^2\theta)\cdot d\theta$$

$$= a^2\int_0^\pi (\tfrac{3}{2} + 2\cos\theta + \tfrac{1}{2}\cos 2\theta)\, d\theta$$

$$= a^2\left\{\tfrac{3}{2}\theta + 2\sin\theta + \tfrac{1}{4}\sin 2\theta\right\}_0^\pi = \tfrac{3}{2}\pi a^2.$$

PROBLEMS

1. Find the area of the lemniscate $r^2 = a^2\cdot\cos 2\theta$.
2. Find the area of $r = 2a\cdot\sin\theta$.
3. Find the length of $r = a\cdot\sin^3\dfrac{\theta}{3}$.
4. Find the area of one loop of $r = a\cdot\sin(2\theta)$.
5. Find the area of $r = a\cdot\sec^2\dfrac{\theta}{2}$ from $\theta = 0$ to $\theta = \dfrac{\pi}{2}$.

Art. 41. — Volume of a Solid by Polar Space Coordinates

The polar space coordinates of a point are r, ϕ, and θ. The conical surfaces corresponding to θ and $\theta + \Delta\theta$ include a conical wedge of the volume to be determined. The planes corresponding to ϕ and $\phi + \Delta\phi$ cut from this wedge a solid of the nature of a pyramid. The spherical surfaces corresponding to r and $r + \Delta r$ cut from this pyramid an element of solid which, when Δr, $\Delta\phi$, and $\Delta\theta$ are indefinitely decreased, approaches as its limit the rectangular parallelopiped whose dimensions are $ab = \Delta r$, $ad = r \cdot \cos\theta \cdot \Delta\phi$, $ac = r \cdot \Delta\theta$. Hence in the limit the element of volume $= r^2 \cdot \cos\theta \cdot dr \cdot d\phi \cdot d\theta$,

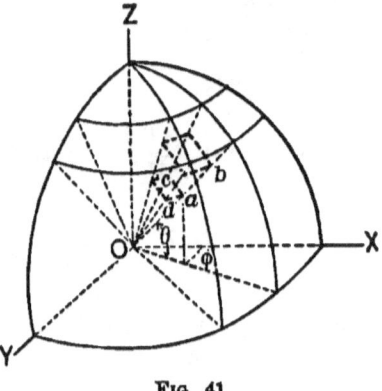

Fig. 41.

the pyramid $= \int_{r=0}^{r=r_1} r^2 \cdot \cos\theta \cdot dr \cdot d\phi \cdot d\theta$,

the wedge $= \int_{\phi=0}^{\phi=\phi_1} \int_{r=0}^{r=r_1} r^2 \cdot \cos\theta \cdot dr \cdot d\phi \cdot d\theta$,

and the entire volume

$$= V = \int_{\theta=0}^{\theta=\theta_1} \int_{\phi=0}^{\phi=\phi_1} \int_{r=0}^{r=r_1} r^2 \cdot \cos\theta \cdot dr \cdot d\phi \cdot d\theta.$$

For example, let it be required to find the volume of the trirectangular spherical pyramid, the radius of the sphere being a. In this problem r extends from 0 to a, θ from 0 to $\frac{\pi}{2}$, ϕ from 0 to $\frac{\pi}{2}$. Hence,

$$V = \int_{\theta=0}^{\theta=\frac{\pi}{2}} \int_{\phi=0}^{\phi=\frac{\pi}{2}} \int_{r=0}^{r=a} r^2 \cdot \cos\theta \cdot dr \cdot d\phi \cdot d\theta$$

$$= \frac{a^3}{3} \int_{\theta=0}^{\theta=\frac{\pi}{2}} \int_{\phi=0}^{\phi=\frac{\pi}{2}} \cos\theta \cdot d\phi \cdot d\theta$$

$$= \frac{\pi a^3}{6} \int_{\theta=0}^{\theta=\frac{\pi}{2}} \cos\theta \cdot d\theta = \frac{\pi a^3}{6}.$$

ART. 42. — DIFFERENTIATION OF INVERSE CIRCULAR FUNCTIONS

To differentiate $y = \sin^{-1} u$, where u is a continuous function of x, write $\sin y = u$, and differentiate with respect to x. There results $\cos y \cdot \dfrac{dy}{dx} = \dfrac{du}{dx}$, whence $\dfrac{dy}{dx} = \dfrac{\dfrac{du}{dx}}{\cos y}$; and, since

$\cos y = \sqrt{1 - \sin^2 y} = \sqrt{1 - u^2}$, $\dfrac{dy}{dx} = \dfrac{d}{dx} \sin^{-1} u = \dfrac{\dfrac{du}{dx}}{\sqrt{1 - u^2}}$.

In like manner $\dfrac{d}{dx} \cos^{-1} u = -\dfrac{\dfrac{du}{dx}}{\sqrt{1 - u^2}}$; $\dfrac{d}{dx} \tan^{-1} u = \dfrac{\dfrac{du}{dx}}{1 + u^2}$;

$\dfrac{d}{dx} \cot^{-1} u = -\dfrac{\dfrac{du}{dx}}{1 + u^2}$; $\dfrac{d}{dx} \sec^{-1} u = \dfrac{\dfrac{du}{dx}}{u \sqrt{u^2 - 1}}$;

$\dfrac{d}{dx} \csc^{-1} u = \dfrac{-\dfrac{du}{dx}}{u \sqrt{u^2 - 1}}$; $\dfrac{d}{dx} \text{vers}^{-1} u = \dfrac{\dfrac{du}{dx}}{\sqrt{2u - u^2}}$;

$\dfrac{d}{dx} \text{covers}^{-1} u = \dfrac{-\dfrac{du}{dx}}{\sqrt{2u - u^2}}$.

CIRCULAR FUNCTIONS

EXAMPLE. — Differentiate $\tan^{-1}\dfrac{2x}{1-x^2}$.

$$\frac{d}{dx}\tan^{-1}\frac{2x}{1-x^2} = \frac{\dfrac{d}{dx}\dfrac{2x}{1-x^2}}{1+\left(\dfrac{2x}{1-x^2}\right)^2} = \frac{\dfrac{(1-x^2)\cdot\dfrac{d}{dx}(2x)-2x\cdot\dfrac{d}{dx}(1-x^2)}{(1-x^2)^2}}{1+\dfrac{4x^2}{(1-x^2)^2}}$$

$$= \frac{2(1-x^2)+4x^2}{(1-x^2)^2+4x^2} = \frac{2(1+x^2)}{1+2x^2+x^4} = \frac{2}{1+x^2}.$$

It appears that $\dfrac{d}{dx}\tan^{-1}\dfrac{2x}{1-x^2} = 2\dfrac{d}{dx}\tan^{-1}x$, which is as it ought to be, since $\tan^{-1}\dfrac{2x}{1-x^2} = 2\tan^{-1}x$ by trigonometry.

PROBLEMS

Differentiate,

1. $\sin^{-1}(3x)$.
2. $3\sin^{-1}\dfrac{x}{2}$.
3. $\sin^{-1}(3x+5)$.
4. $\tan^{-1}\dfrac{2x}{3}$.
5. $5\operatorname{vers}^{-1}\dfrac{x}{5}$.
6. $3x^2+2\cos^{-1}x$.
7. $\sec^{-1}(x^2)$.
8. $\cot^{-1}\left(\dfrac{1}{x}\right)$.
9. $5\tan^{-1}\dfrac{2}{x}$.
10. $\cot^{-1}\dfrac{1-x}{1+x}$.

11. Show that $\dfrac{d}{dx}\sin^{-1}(3x-4x^3) = 3\dfrac{d}{dx}\sin^{-1}x$.

12. Form $\dfrac{dy}{dx}$ and $\dfrac{d^2y}{dx^2}$ of equation of cycloid

$$x = r\operatorname{vers}^{-1}\dfrac{y}{r} - \sqrt{2ry-y^2}.$$

Here $\dfrac{dx}{dy} = \dfrac{y}{\sqrt{2ry-y^2}}$; hence $\dfrac{dy}{dx} = \sqrt{\dfrac{2r-y}{y}}$.

Squaring, $\dfrac{dy^2}{dx^2} = \dfrac{2r}{y}-1$. Differentiating, $2\dfrac{dy}{dx}\dfrac{d^2y}{dx^2} = -\dfrac{2r}{y^2}\dfrac{dy}{dx}$.

Whence, $\dfrac{d^2y}{dx^2} = -\dfrac{r}{y^2}$.

13. Form $\dfrac{dy}{dx}$ and $\dfrac{d^2y}{dx^2}$ of cycloid $y = r \cdot \text{vers}^{-1}\dfrac{x}{r} + \sqrt{2rx - x^2}$.

14. Find length of cycloid $y = r \cdot \text{vers}^{-1}\dfrac{x}{r} + \sqrt{2rx - x^2}$.

15. On a pedestal 25 feet high stands a statue 11 feet high. Find the distance from the base of the pedestal of the point in the horizontal plane through the base at which the statue subtends the greatest angle.

ART. 43. — INTEGRATION BY INVERSE CIRCULAR FUNCTIONS

The results of the preceding article when $\dfrac{u}{a}$ is substituted for u may be written in the form,

I. $\displaystyle\int \dfrac{\dfrac{du}{dx}}{\sqrt{a^2 - u^2}} = \sin^{-1}\dfrac{u}{a} + C;$ III. $\displaystyle\int \dfrac{\dfrac{du}{dx}}{a^2 + u^2} = \dfrac{1}{a}\tan^{-1}\dfrac{u}{a} + C;$

II. $\displaystyle\int \dfrac{-\dfrac{du}{dx}}{\sqrt{a^2 - u^2}} = \cos^{-1}\dfrac{u}{a} + C;$ IV. $\displaystyle\int \dfrac{-\dfrac{du}{dx}}{a^2 + u^2} = \dfrac{1}{a}\cot^{-1}\dfrac{u}{a} + C;$

V. $\displaystyle\int \dfrac{\dfrac{du}{dx}}{u\sqrt{u^2 - a^2}} = \dfrac{1}{a}\sec^{-1}\dfrac{u}{a} + C;$

VI. $\displaystyle\int \dfrac{-\dfrac{du}{dx}}{u\sqrt{u^2 - a^2}} = \dfrac{1}{a}\text{cosec}^{-1}\dfrac{u}{a} + C;$

VII. $\displaystyle\int \dfrac{\dfrac{du}{dx}}{\sqrt{2au - u^2}} = \text{vers}^{-1}\dfrac{u}{a} + C;$

VIII. $\displaystyle\int \dfrac{-\dfrac{du}{dx}}{\sqrt{2au - u^2}} = \text{covers}^{-1}\dfrac{u}{a} + C.$

CIRCULAR FUNCTIONS

EXAMPLE. — Integrate $\dfrac{dy}{dx} = \dfrac{x^{\frac{1}{2}}}{\sqrt{2-4x^3}}$. This derivative has the general form $\dfrac{\frac{du}{dx}}{\sqrt{a^2-u^2}}$. Placing $u^2 = 4x^3$, $u = 2x^{\frac{3}{2}}$ and $\dfrac{du}{dx} = 3x^{\frac{1}{2}}$. Writing $\dfrac{dy}{dx} = \dfrac{x^{\frac{1}{2}}}{\sqrt{2-4x^3}} = \dfrac{1}{3}\dfrac{\frac{d}{dx}(2x^{\frac{3}{2}})}{\sqrt{2-4x^3}}$,

$$y = \tfrac{1}{3}\sin^{-1}(\sqrt{2}\cdot x^{\frac{3}{2}}) + C.$$

PROBLEMS

Integrate,

1. $\dfrac{dy}{dx} = \dfrac{3}{4+9x^2}.$

2. $\dfrac{dy}{dx} = \dfrac{2}{x\sqrt{3x^2-5}}.$

3. $\dfrac{dy}{dx} = \dfrac{x}{1+x^4}.$

4. $\dfrac{dy}{dx} = \dfrac{3}{\sqrt{5x^4-2x^2}}.$

5. $\dfrac{dy}{dx} = \dfrac{x^4}{1+x^2}.$ Reduce improper fraction to mixed number.

6. $\dfrac{dy}{dx} = \dfrac{x}{x^4+4}.$

7. $\dfrac{dy}{dx} = \dfrac{1}{\sqrt{6x-x^2}}.$

8. $\dfrac{dy}{dx} = \dfrac{x}{\sqrt{1-x^4}}.$

9. $\dfrac{dy}{dx} = \dfrac{\tan^{-1}x}{1+x^2}.$

10. $\dfrac{dy}{dx} = \dfrac{\sin^{-1}x}{\sqrt{1-x^2}}.$

11. $\dfrac{dy}{dx} = \dfrac{\sec^{-1}x}{x\sqrt{x^2-1}}.$

12. $\dfrac{dy}{dx} = \dfrac{1}{x^2-6x+11}.$ Since $\dfrac{dy}{dx} = \dfrac{1}{2+(x-3)^2} = \dfrac{\frac{d}{dx}(x-3)}{2+(x-3)^2}$,

$y = \dfrac{1}{\sqrt{2}}\tan^{-1}\dfrac{x-3}{\sqrt{2}} + C.$

13. $\dfrac{dy}{dx} = \dfrac{1}{\sqrt{1+3x-x^2}}.$

14. The time of descent of a body down the arc of the vertical frictionless cycloid $x = r \cdot \text{vers}^{-1}\frac{y}{r} + \sqrt{2ry - y^2}$, from

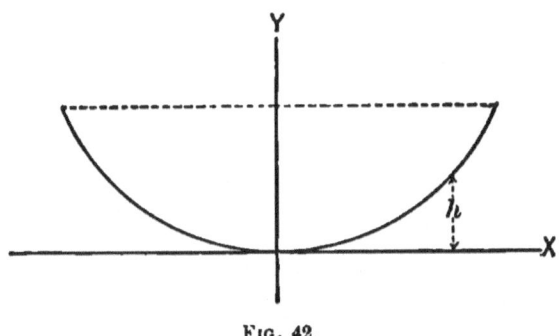

Fig. 42.

$y = h$ to the vertex, is $t = \left(\frac{r}{g}\right)^{\frac{1}{2}} \int_0^h \frac{dy}{\sqrt{hy - y^2}}$. Show that the time is the same for all positions of the starting point.

15. A body is suspended on a frictionless horizonal axis, turned through a small angle θ_0, and then left free under the action of gravity. This constitutes a compound pendulum. The relation between θ, the angle the pendulum in any position makes with the vertical and the time t measured in seconds after the pendulum is started, is expressed by the equation $\frac{d^2\theta}{dt^2} = \frac{-gh}{h^2 + k_1^2}\theta$, where g is the acceleration of gravity, h and k_1 are constants depending on the shape and material of the pendulum and the position of the axis. Find the time of a complete oscillation. In this problem, when $t = 0$, $\theta = \theta_0$, $\frac{d\theta}{dt} = 0$.

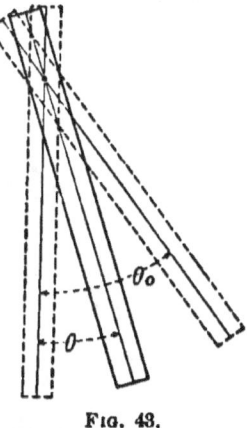

Fig. 43.

16. In strength of materials it is proved that for any point (x, y) of the elastic curve of a long column

$$EI\frac{d^2y}{dx^2} = -P \cdot y,$$

where E and I are constants depending on the material and cross-section of the column; P is the load. Calling the maximum deflection Δ, when $y = \Delta$, $\frac{dy}{dx} = 0$, and when $x = 0$, $y = 0$. Find the equation of the elastic curve.

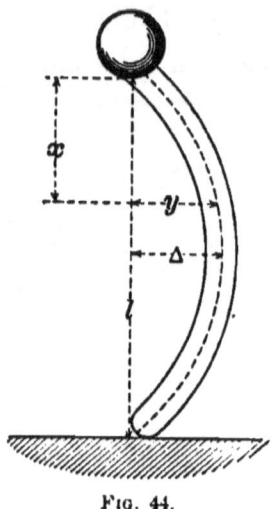

Fig. 44.

Art. 44. — Radius of Curvature

If a point moves in the circumference of a circle, the tangent to the circle at this point changes its direction. Suppose the point to start from A. When it reaches B, the tangent has turned through the angle $T'ST = AOB$. The ratio of the angle AOB to the distance AB the point has moved along the circle is called the rate of curvature of the circle, and equals $\frac{\theta}{r \cdot \theta} = \frac{1}{r}$. That is, the rate of curvature of the circle is the reciprocal of the radius of the circle.

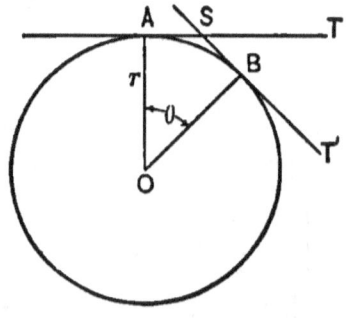

Fig. 45.

If a point moves along any curve $y = f(x)$, the ratio of the angle through which the tangent turns to the distance the point moves is called the average rate of curvature of the curve for the distance the point moves. Denoting by $\Delta\phi$

the angle through which the tangent turns while the point moves a distance Δs along the curve, the average rate of curvature is $\dfrac{\Delta \phi}{\Delta s}$. The actual rate of curvature at any point (x, y) of the curve is the value at (x, y) of $\lim \dfrac{\Delta \phi}{\Delta s} = \dfrac{d\phi}{ds}$ when limit $\Delta s = 0$.

The reciprocal of the rate of curvature at any point (x, y) of the curve is the radius of the circle which has the same rate of curvature as the curve at the point (x, y). The reciprocal of the rate of curvature is called the radius of curvature, and is denoted by ρ, so that $\rho = \dfrac{ds}{d\phi}$. Now $\phi = \tan^{-1} \dfrac{dy}{dx}$, hence

$$\rho = \frac{ds}{d\phi} = \frac{ds}{dx} \cdot \frac{dx}{d\phi} = \frac{\dfrac{ds}{dx}}{\dfrac{d\phi}{dx}} = \frac{\left(1 + \dfrac{dy^2}{dx^2}\right)^{\frac{1}{2}}}{\dfrac{\dfrac{d^2y}{dx^2}}{1 + \dfrac{dy^2}{dx^2}}} = \frac{\left(1 + \dfrac{dy^2}{dx^2}\right)^{\frac{3}{2}}}{\dfrac{d^2y}{dx^2}}.$$

The analysis supposes that the curve is such that y is a continuous function of x, and ϕ a continuous function of s.

If the equation of the curve is given in the form $x = f_1(t)$, $y = f_2(t)$, $\dfrac{dy}{dx} = \dfrac{\dfrac{dy}{dt}}{\dfrac{dx}{dt}}$, and

$$\frac{d^2y}{dx^2} = \frac{d}{dx} \frac{\dfrac{dy}{dt}}{\dfrac{dx}{dt}} = \frac{d}{dt} \frac{\dfrac{dy}{dt}}{\dfrac{dx}{dt}} \cdot \frac{1}{\dfrac{dx}{dt}} = \frac{\dfrac{d^2y}{dt^2} \cdot \dfrac{dx}{dt} - \dfrac{d^2x}{dt^2} \cdot \dfrac{dy}{dt}}{\dfrac{dx^3}{dt^3}}.$$

Hence $$\rho = \frac{\left(\dfrac{dx^2}{dt^2} + \dfrac{dy^2}{dt^2}\right)^{\frac{3}{2}}}{\dfrac{d^2y}{dt^2} \cdot \dfrac{dx}{dt} - \dfrac{d^2x}{dt^2} \cdot \dfrac{dy}{dt}}.$$

For a polar curve $r = f(\theta)$, $x = r \cdot \cos \theta$, $y = r \cdot \sin \theta$. Hence

$$\frac{dx}{d\theta} = -r \cdot \sin \theta + \cos \theta \cdot \frac{dr}{d\theta},$$

$$\frac{d^2x}{d\theta^2} = -r \cdot \cos \theta - 2 \sin \theta \cdot \frac{dr}{d\theta} + \cos \theta \cdot \frac{d^2r}{d\theta^2},$$

$$\frac{dy}{d\theta} = r \cdot \cos \theta + \sin \theta \cdot \frac{dr}{d\theta},$$

$$\frac{d^2y}{d\theta^2} = -r \cdot \sin \theta + 2 \sin \theta \cdot \frac{dr}{d\theta} + \sin \theta \cdot \frac{d^2r}{d\theta^2}.$$

Hence for a polar curve $\rho = \dfrac{\left(r^2 + \dfrac{dr^2}{d\theta^2}\right)^{\frac{3}{2}}}{r^2 + 2\dfrac{dr^2}{d\theta^2} - r^2 \cdot \dfrac{d^2r}{d\theta^2}}.$

A circle of radius ρ, placed so that the circle and curve $y = f(x)$ have a common tangent at (x, y) and lie on the same side of the common tangent, is called the circle of curvature of $y = f(x)$ at (x, y); the center of the circle is called the center of curvature of $y = f(x)$ at (x, y). Denoting the coordinates of the center of curvature by α and β,

(1) $\alpha = x - \rho \cdot \sin \phi$

$= x - \dfrac{\dfrac{dy}{dx}\left(1 + \dfrac{dy^2}{dx^2}\right)}{\dfrac{d^2y}{dx^2}};$

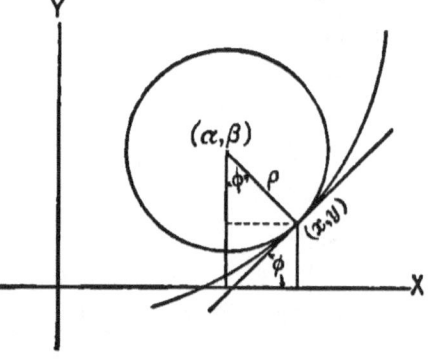

Fig. 46.

(2) $\beta = y + \rho \cdot \cos \phi = y + \dfrac{1 + \dfrac{dy^2}{dx^2}}{\dfrac{d^2y}{dx^2}}.$

The locus of the center of curvature as the point (x, y) traces the curve $y = f(x)$ is called the evolute of $y = f(x)$. The equation of the evolute is found by solving (1) and (2) for x and y in terms of α and β and substituting in $y = f(x)$.

EXAMPLE I. — Find the radius of curvature, coordinates of center of curvature, and evolute of the parabola $y^2 = 2px$.

Here $\dfrac{dy}{dx} = \dfrac{p}{y}$, $\dfrac{d^2y}{dx^2} = -\dfrac{p}{y^3}$, hence $\rho = -\dfrac{(p^2+y^2)^{\frac{3}{2}}}{p^2}$; $\alpha = 3x+p$, $\beta = -\dfrac{y^3}{p^2}$; the evolute is $\beta^2 = \frac{8}{27}(\alpha-p)^3$, a semi-cubic parabola.

The radius of curvature to the parabola at the vertex $(0, 0)$ is p; at the extremity of the latus rectum $(\frac{1}{2}p, p)$ the radius of curvature is $2\sqrt{2} \cdot p$.

EXAMPLE II. — Find the radius of curvature of $xy = 4$ at the point $(1, 4)$.

For this curve $\dfrac{dy}{dx} = -\dfrac{4}{x^2}$, $\dfrac{d^2y}{dx^2} = \dfrac{8}{x^3}$. At $(1, 4)$, $\dfrac{dy}{dx} = -4$, $\dfrac{d^2y}{dx^2} = 8$. Hence $\rho = \frac{1}{8}(2)^{\frac{3}{2}}$.

PROBLEMS

Find the radius of curvature of

1. $y^2 = 4x$ at $(1, 2)$; $(0, 0)$; $(4, 4)$.
2. $\dfrac{x^2}{9} + \dfrac{y^2}{4} = 1$ at $(3, 0)$; $(0, 2)$.
3. $\dfrac{x^2}{a^2} + \dfrac{y^2}{b^2} = 1$.
4. $\dfrac{x^2}{a^2} - \dfrac{y^2}{b^2} = 1$.
5. $x = r \cdot \text{vers}^{-1}\dfrac{y}{r} - \sqrt{2ry - y^2}$.
6. $x = v \cdot t$, $y = \frac{1}{2}g \cdot t^2$.
7. $x = a\cos\phi$, $y = b\sin\phi$.
8. $x = R \cdot \theta - R \cdot \sin\theta$, $y = R - R \cdot \cos\theta$.

9. $2xy = a^2$.
10. $x^{\frac{2}{3}} + y^{\frac{2}{3}} = a^{\frac{2}{3}}$.
11. $y = \sin x$.
12. $r = a(1 - \cos \theta)$.
13. $r^2 = a^2 \cdot \cos(2\theta)$.
14. $r = a^\theta$.

CHAPTER VIII

LOGARITHMIC AND EXPONENTIAL FUNCTIONS

ART. 45. — The Limit of $\left(1+\dfrac{1}{z}\right)^z$ When Limit $z = \infty$

Assume first that z takes only positive integral values m. By the binomial formula,

$$\left(1+\frac{1}{m}\right)^m \equiv 1 + m\cdot\frac{1}{m} + \frac{m(m-1)}{1\cdot 2}\cdot\frac{1}{m^2}$$

$$+ \frac{m(m-1)(m-2)}{1\cdot 2\cdot 3}\frac{1}{m^3} + \cdots$$

$$+ \frac{m(m-1)(m-2)\cdots(m-n+1)}{1\cdot 2\cdot 3\cdots n}\frac{1}{m^n}$$

$$+ \frac{m(m-1)(m-2)\cdots(m-n+1)(m-n)}{1\cdot 2\cdot 3\cdot 4\cdots n\cdot(n+1)}\frac{1}{m^{n+1}} + \cdots$$

$$+ \frac{m(m-1)(m-2)\cdots\{m-(m-1)\}}{1\cdot 2\cdot 3\cdot 4\cdots m}\frac{1}{m^m}$$

$$\equiv 1 + 1 + \frac{1-\dfrac{1}{m}}{1\cdot 2} + \frac{\left(1-\dfrac{1}{m}\right)\left(1-\dfrac{2}{m}\right)}{1\cdot 2\cdot 3} + \cdots$$

$$+ \frac{\left(1-\dfrac{1}{m}\right)\left(1-\dfrac{2}{m}\right)\cdots\left(1-\dfrac{n-1}{m}\right)}{1\cdot 2\cdot 3\cdots n}$$

LOGARITHMIC AND EXPONENTIAL FUNCTIONS

$$+\frac{\left(1-\frac{1}{m}\right)\left(1-\frac{2}{m}\right)\cdots\left(1-\frac{n-1}{m}\right)}{1\cdot 2\cdot 3\cdots n}\left\{1-\frac{n}{m}+\frac{\left(1-\frac{n}{m}\right)\left(1-\frac{n+1}{m}\right)}{(n+1)(n+2)}+\cdots\right.$$

$$\left.+\frac{\left(1-\frac{n}{m}\right)\left(1-\frac{n+1}{m}\right)\cdots\left(1-\frac{m-1}{m}\right)}{(n+1)(n+2)\cdots m}\right\}.$$

This expansion is true for positive integral values of m however large m may be taken. Denote by S the sum of the first $n+1$ terms of the expansion, by R the sum of the remaining terms. Then the equation

$$\text{limit}\left(1+\frac{1}{z}\right)^z = \text{limit } S + \text{limit } R$$

is always true. When m is indefinitely increased,

$$\text{limit } S = 1 + 1 + \frac{1}{1\cdot 2} + \frac{1}{1\cdot 2\cdot 3} + \frac{1}{1\cdot 2\cdot 3\cdot 4} + \cdots \frac{1}{1\cdot 2\cdot 3\cdot 4\cdots n}.$$

Since $R < \dfrac{1}{1\cdot 2\cdot 3\cdots n}\left\{\dfrac{1}{n+1} + \dfrac{1}{(n+1)^2} + \dfrac{1}{(n+1)^3} + \cdots \right.$

$$\left. + \frac{1}{(n+1)^m}\right\},$$

when m is indefinitely increased, $\text{limit } R < \dfrac{1}{n(1\cdot 2\cdot 3\cdots n)}$. Now n may be taken so large that limit R becomes less than any quantity that can be assigned. Hence, when limit $m=\infty$,

$$\text{limit}\left(1+\frac{1}{m}\right)^m$$

$$= \text{limit}\left\{1+1+\frac{1}{1\cdot 2}+\frac{1}{1\cdot 2\cdot 3}+\frac{1}{1\cdot 2\cdot 3\cdot 4}+\cdots \frac{1}{1\cdot 2\cdot 3\cdot 4\cdots n}\right\}$$

when limit $n=\infty$.

The value of limit $\left(1+\dfrac{1}{m}\right)^m$ when limit $m = \infty$ to nine decimals is 2.718281828. This limit is the base of the Napierian system of logarithms and is denoted by e.

Next assume that z may take fractional as well as integral positive values as it approaches infinity. Denote any one value of z by s, and let m be an integer such that $m < s \leqq m + 1$. Then $\left(1+\dfrac{1}{m+1}\right)^m < \left(1+\dfrac{1}{s}\right)^s < \left(1+\dfrac{1}{m}\right)^{m+1}$, which may be written $\dfrac{\left(1+\dfrac{1}{m+1}\right)^{m+1}}{1+\dfrac{1}{m+1}} < \left(1+\dfrac{1}{s}\right)^s < \left(1+\dfrac{1}{m}\right)^m \cdot \left(1+\dfrac{1}{m}\right).$

When limit $s = \infty$, limit $m = \infty$. Hence when limit $s = \infty$, limit $\left(1+\dfrac{1}{s}\right)^s$ lies between two quantities whose common limit is e. Consequently limit $\left(1+\dfrac{1}{s}\right)^s = e$ when limit $s = \infty$.

Lastly assume that $z = -r$, and that limit $r = \infty$. Now

$$\text{limit}\left(1-\dfrac{1}{r}\right)^{-r} = \text{limit}\left(\dfrac{r-1}{r}\right)^{-r} = \text{limit}\left(\dfrac{r}{r-1}\right)^r$$

$$= \text{limit}\left(1+\dfrac{1}{r-1}\right)^r$$

$$= \text{limit}\left(1+\dfrac{1}{r-1}\right)^{r-1} \cdot \left(1+\dfrac{1}{r-1}\right) = e$$

when limit $r = \infty$.

It appears that in every case, when limit $z = \infty$, limit $\left(1+\dfrac{1}{z}\right)^z = e = 2.718281828$. This limit is fundamental in this chapter.

LOGARITHMIC AND EXPONENTIAL FUNCTIONS

ART. 46. — DIFFERENTIATION OF LOGARITHMIC FUNCTIONS

Let u represent a continuous function of x, and denote by Δu and Δx corresponding changes in u and x. Then, if a is the base of the system of logarithms used,

$$\frac{d}{dx}\log_a u = \text{limit } \frac{\log_a(u + \Delta u) - \log_a u}{\Delta u} \cdot \frac{\Delta u}{\Delta x}$$

$$= \text{limit } \frac{1}{u} \cdot \frac{\log_a \frac{u + \Delta u}{u}}{\frac{\Delta u}{u}} \cdot \frac{\Delta u}{\Delta x}$$

$$= \text{limit } \frac{1}{u} \cdot \log_a\left(1 + \frac{\Delta u}{u}\right)^{\frac{1}{\Delta u}} \cdot \frac{\Delta u}{\Delta x}$$

$$= \text{limit } \frac{1}{u} \cdot \log_a\left[1 + \frac{1}{\frac{u}{\Delta u}}\right]^{\frac{u}{\Delta u}} \cdot \frac{\Delta u}{\Delta u}$$

$$= \log_a e \cdot \frac{\frac{du}{dx}}{u} \quad \text{when limit } \Delta x = 0.$$

For by the nature of logarithms the difference of the logarithms of two numbers is the logarithm of the quotient of the numbers, and the logarithm of a number affected by an exponent is the exponent times the logarithm of the number. Since u is a continuous function of x, limit $\Delta u = 0$ when limit $\Delta x = 0$. Writing $\frac{u}{\Delta u} = z$, when limit $\Delta u = 0$, limit $z = \infty$. Hence limit $\left[1 + \frac{1}{\frac{u}{\Delta u}}\right]^{\frac{u}{\Delta u}}$ when limit $\Delta x = 0$ equals limit $\left(1 + \frac{1}{z}\right)^z$ when limit $z = \infty$.

Calling, as is customary, $\log_a e$ the modulus of the system of logarithms whose base is a, and denoting it by M,

$\dfrac{d}{dx}\log_a u = M\dfrac{\dfrac{du}{dx}}{u}$; in words, the derivative of the logarithm of a function of x is the modulus of the system of logarithms times the derivative of the function divided by the function.

In the Napierian system of logarithms $\dfrac{d}{dx}\log_e u = \dfrac{1}{u} \cdot \dfrac{du}{dx}$. Unless otherwise specified, the Napierian system is to be used.

EXAMPLE. — Differentiate $\log\sqrt{\dfrac{1-x}{1+x}}$.

$$\dfrac{d}{dx}\log\sqrt{\dfrac{1-x}{1+x}} = \dfrac{d}{dx}\tfrac{1}{2}\{\log(1-x) - \log(1+x)\}$$

$$= \tfrac{1}{2}\left\{\dfrac{-1}{1-x} - \dfrac{1}{1+x}\right\} = \dfrac{-1}{1-x^2}.$$

PROBLEMS

Differentiate,

1. $\log x^2$.
2. $\log\dfrac{1}{x}$.
3. $\log(3x - 5)$.
4. $\log(x^{\frac{1}{2}})$.
5. $\log^2 x$.
6. $\log \sin x$.
7. $\log \tan x$.
8. $\log \sqrt{1 - x^2}$.
9. $\log\dfrac{1-x^2}{1+x^2}$.
10. $\log \tan \tfrac{1}{2} x$.
11. $\log\sqrt{\dfrac{1-\cos x}{1+\cos x}}$.
12. $\log \dfrac{1}{\cos x}$.
13. $\log \dfrac{x}{\sqrt{1+x^2}}$.
14. $\log(x + \sqrt{1+x^2})$.
15. $\log(x + \sqrt{x^2 + a^2})$.
16. $\log(x + \sqrt{x^2 - a^2})$.

Find the true value of,

17. $\dfrac{\log x}{(1-x)^{\frac{1}{2}}}$ when $x = 1$.
18. $\dfrac{\log x}{x - 1}$ when $x = 1$.

LOGARITHMIC AND EXPONENTIAL FUNCTIONS 133

19. $\dfrac{\log x}{x^n}$ when $x = \infty$.

20. $\dfrac{\log x}{\cot x}$ when $x = 0$.

21. $\dfrac{\log \tan (2x)}{\log \tan x}$ when $x = 0$.

22. $x^m \log^n x$ when $x = 0$.

23. $\dfrac{x}{x-1} - \dfrac{1}{\log x}$ when $x = 1$.

24. $\dfrac{\log \sin x}{(\pi - 2x)^2}$ when $x = \dfrac{\pi}{2}$.

ART. 47. — INTEGRATION BY LOGARITHMIC FUNCTIONS

The result of the preceding article may be written

$$\int \dfrac{\dfrac{du}{dx}}{u} = \int u^{-1} \cdot \dfrac{du}{dx} = \log_e u + C.$$

EXAMPLE. — Integrate $\dfrac{dy}{dx} = \dfrac{x}{1+x^2}$.

The first derivative of the denominator is $2x$, hence

$$\dfrac{dy}{dx} = \dfrac{1}{2} \dfrac{\dfrac{d}{dx}(1+x^2)}{1+x^2} \quad \text{and} \quad y = \dfrac{1}{2}\log_e (1+x^2) + C.$$

PROBLEMS

Integrate,

1. $\dfrac{dy}{dx} = \dfrac{1+3x}{2x+3x^2}$.

2. $\dfrac{dy}{dx} = \dfrac{3x^4}{x^5+7}$.

3. $\dfrac{dy}{dx} = \dfrac{nx\,dx}{a^2+x^2}$.

4. $\dfrac{dy}{dx} = \dfrac{2x+3}{x^2+3x}$.

5. $\dfrac{dy}{dx} = \tan x$.

6. $\dfrac{dy}{dx} = \cot x$.

7. $\dfrac{dy}{dx} = \dfrac{1}{\sin x}$. Writing $\dfrac{dy}{dx} = \dfrac{1}{\sin x} = \dfrac{1}{2 \sin \tfrac{1}{2}x \cdot \cos \tfrac{1}{2}x}$

$= \dfrac{\tfrac{1}{2} \sec^2 \tfrac{1}{2}x}{\tan \tfrac{1}{2}x} = \dfrac{\sec^2 \tfrac{1}{2}x \cdot \dfrac{d}{dx}(\tfrac{1}{2}x)}{\tan \tfrac{1}{2}x} = \dfrac{\dfrac{d}{dx}\tan \tfrac{1}{2}x}{\tan \tfrac{1}{2}x}$, $y = \log \tan \tfrac{1}{2}x + C$.

8. $\dfrac{dy}{dx} = \dfrac{1}{\cos x}$.   9. $\dfrac{dy}{dx} = \dfrac{1}{x(1+x^2)}$. Substitute $x = \tan\theta$.

10. $\dfrac{dy}{dx} = \dfrac{1}{\sqrt{x^2+a^2}}$. Write $\sqrt{x^2+a^2} = z - x$, whence $z = x + \sqrt{x^2+a^2}$, $\dfrac{dx}{dz} = \dfrac{z-x}{z}$, $\dfrac{dy}{dz} = \dfrac{dy}{dx} \cdot \dfrac{dx}{dz} = \dfrac{1}{z-x} \cdot \dfrac{z-x}{z} = \dfrac{1}{z}$, and $y = \log_e z + C = \log_e (x + \sqrt{x^2+a^2}) + C$.

11. $\dfrac{dy}{dx} = \dfrac{1}{\sqrt{x^2-a^2}}$.   12. $\dfrac{dy}{dx} = \dfrac{1}{\sqrt{x^2+2x}}$.

Writing $\dfrac{1}{\sqrt{x^2+2x}} = \dfrac{1}{\sqrt{(x+1)^2-1}}$, $\dfrac{dy}{dx} = \dfrac{\frac{d}{dx}(x+1)}{\sqrt{(x+1)^2-1}}$, and $y = \log\{(x+1) + \sqrt{x^2+2x}\} + C$.

13. $\dfrac{dy}{dx} = \dfrac{2+3x}{1+x^2}$. Write $\dfrac{dy}{dx} = \dfrac{2}{1+x^2} + \dfrac{3x}{1+x^2}$ and integrate term by term.

14. $\dfrac{dy}{dx} = \dfrac{m+nx}{a^2+x^2}$.   15. $\dfrac{dy}{dx} = \dfrac{\log^2 x}{x}$.   16. $\dfrac{dy}{dx} = \dfrac{\log^n x}{x}$.

17. $\dfrac{dy}{dx} = \dfrac{x^3}{x^2+1}$. Reduce $\dfrac{x^3}{x^2+1}$ to a mixed number.

18. Find the area of $xy = 1$ from $x = 0$ to $x = x'$.

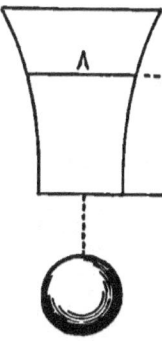

Fig. 47.

19. If A is the cross-section of a bar of uniform strength at a distance y from the lower end, $\dfrac{dA}{dy} = \dfrac{w \cdot A}{S}$, where w is the weight of the bar per cubic foot, and S is a constant depending on the material of the bar and the area of the lower end. Find the relation between A and y.

Art. 48. — Integration by Partial Fractions

The function of x defined by $\dfrac{F(x)}{\phi(x)}$ is called a rational fraction when x does not occur affected by a fractional exponent or under a radical sign.

If the numerator of the rational fraction $\dfrac{F(x)}{\phi(x)}$ is not of lower degree than the denominator, the given fraction may always be transformed by division into the sum of an integral function $P(x)$ and a rational fraction $\dfrac{f(x)}{\phi(x)}$, whose numerator is of lower degree than its denominator. For example,

$$\frac{x^4}{x^2+1} \equiv x^2 - 1 + \frac{1}{x^2+1}.$$

Consider the rational fraction $\dfrac{f(x)}{\phi(x)}$ whose numerator is of lower degree in x than the denominator. Suppose the denominator $\phi(x)$ to be of degree n, then $f(x)$ cannot be of higher degree than $n-1$. By a theorem in algebra $\phi(x)$ can be resolved into n factors of the first degree, and imaginary factors must occur in conjugate pairs. The product of a pair of conjugate imaginary factors of the first degree,

$$(x - c - d\sqrt{-1})(x - c + d\sqrt{-1}) = (x-c)^2 + d^2,$$

a real factor of the second degree. Hence $\phi(x)$ can be resolved into real factors of the first and second degrees.

Let $\phi(x) \equiv (x-a)(x-b)^s\{(x-c)^2+d^2\}\{(x-e)^2+f^2\}^{t\cdots}$. It is proposed to break up the fraction $\dfrac{f(x)}{\phi(x)}$ into the sum of partial fractions whose denominators are the factors of $\phi(x)$, and in every partial fraction the numerator is to be of lower degree than the denominator.

Assume

$$\frac{f(x)}{\phi(x)} \equiv \frac{A}{x-a} + \frac{B_1}{(x-b)^s} + \frac{B_2}{(x-b)^{s-1}} + \cdots \frac{B_s}{x-b} + \frac{Cx+D}{(x-c)^2+d^2}$$

$$+ \frac{E_1 x + F_1}{\{(x-e)^2+f^2\}^t} + \frac{E_2 x + F_2}{\{(x-e)^2+f^2\}^{t-1}} + \cdots \frac{E_t x + F_t}{(x-e^2)+f^2}.$$

The number of undetermined constants,

$$A,\ B_1,\ B_2,\ \cdots,\ B_s,\ C,\ D,\ E_1,\ F_1,\ E_2,\ F_2,\ \cdots,\ E_t,\ F_t$$

introduced equals the degree of $\phi(x)$. Multiplying both members of the identical equation by $\phi(x)$, the right hand member will be of degree $n-1$ in x, while, by hypothesis, $f(x)$ cannot be of a higher degree than $n-1$. Collecting the terms of like powers of x in the right hand member, the coefficients of the resulting n terms are linear in the n undetermined constants. Hence, by equating the coefficients of corresponding terms of both members of the identity, there result n equations linear in the n assumed constants. These equations determine the assumed constants uniquely. Hence the fraction $\frac{f(x)}{\phi(x)}$ can be broken up into partial fractions of the form assumed and in only one way.

A few examples will explain the process of breaking up a fraction and show the importance of partial fractions in integration.

EXAMPLE I. — Integrate $\frac{dy}{dx} = \frac{x^3-1}{x^2-4}$.

Transforming $\frac{x^3-1}{x^2-4}$ to a mixed number, $\frac{x^3-1}{x^2-4} \equiv x + \frac{4x-1}{x^2-4}$.

Assume $\qquad \frac{4x-1}{x^2-4} \equiv \frac{A}{x+2} + \frac{B}{x-2}$,

whence $\qquad 4x-1 \equiv (A+B)x + (-2A+2B).$

Equating the coefficients of like powers of x,

$$A+B=2, \quad -2A+2B=-1, \text{ and } A=\tfrac{9}{4}, \ B=\tfrac{7}{4}.$$

Hence, $\dfrac{dy}{dx} = \dfrac{x^3-1}{x^2-4} \equiv x + \dfrac{4x-1}{x^2-4} \equiv x + \dfrac{9}{4}\dfrac{1}{x+2} + \dfrac{7}{4}\dfrac{1}{x-2}.$

Integrating, $y = \dfrac{x^2}{2} + \tfrac{9}{4}\log(x+2) + \tfrac{7}{4}\log(x-2) + C.$

EXAMPLE II. — Integrate $\dfrac{dy}{dx} = \dfrac{5x+1}{x^2+x-2}.$

Assume $\dfrac{5x+1}{x^2+x-2} \equiv \dfrac{A}{x-1} + \dfrac{B}{x+2},$

whence $5x+1 \equiv A(x+2) + B(x-1).$

This identity is true for all values of x. When $x=1$, $A=2$; when $x=-2$, $B=3$. Hence, $\dfrac{dy}{dx} = \dfrac{2}{x-1} + \dfrac{3}{x+2}.$ Integrating, $y = 2\log(x-1) + 3\log(x+2) + C.$ This result may also be written,

$$y = \log(x+1)^2 + \log(x+2)^3 + \log c$$
$$= \log\{c(x-1)^2(x+2)^3\}.$$

EXAMPLE III. — Integrate $\dfrac{dy}{dx} = \dfrac{x^2+4x-2}{1+x+x^2+x^3}.$

Assume $\dfrac{x^2+4x-2}{1+x+x^2+x^3} \equiv \dfrac{A}{1+x} + \dfrac{Bx+C}{1+x^2},$

whence $x^2+4x-2 \equiv (A+B)x^2 + (B+C)x + A.$

Equating the coefficients of like powers of x, $A=-2$, $B+C=4$, $A+B=1$, whence $A=-2$, $B=3$, $C=1$.

Hence, $\dfrac{dy}{dx} = \dfrac{-2}{1+x} + \dfrac{3x}{1+x^2} + \dfrac{1}{1+x^2};$

and $y = -2\log(1+x) + \tfrac{3}{2}\log(1+x^2) + \tan^{-1}x + C.$

EXAMPLE IV. — Integrate $\dfrac{dy}{dx} = \dfrac{x^3 + x - 1}{(x^2 + 2)^2}$.

Assume $\quad \dfrac{x^3 + x - 1}{(x^2 + 2)^2} \equiv \dfrac{Ax + B}{(x^2 + 2)^2} + \dfrac{Cx + D}{x^2 + 2}$.

By clearing of fractions and equating the coefficients of like powers of x, it is found that $A = -1$, $B = -1$, $C = 1$, $D = 0$.

Hence, $\dfrac{dy}{dx} = \dfrac{-x-1}{(x^2+2)^2} + \dfrac{x}{x^2+2} = \dfrac{-x}{(x^2+2)^2} + \dfrac{x}{x^2+2} - \dfrac{1}{(x^2+2)^2}$.

The first two terms are directly integrable. To integrate the last term substitute $x = \sqrt{2} \cdot \tan\theta$.

PROBLEMS

Integrate,

1. $\dfrac{dy}{dx} = \dfrac{2x+3}{x^3+x^2-2x}$.

2. $\dfrac{dy}{dx} = \dfrac{a}{x^2-a^2}$.

3. $\dfrac{dy}{dx} = \dfrac{2-3x^2}{(x+2)^2}$.

4. $\dfrac{dy}{dx} = \dfrac{x^2}{x^4+x^2-2}$.

5. $\dfrac{dy}{dx} = \dfrac{1}{x^3-x^2+2x-2}$.

6. $\dfrac{dy}{dx} = \dfrac{1}{x^3+1}$.

7. $\dfrac{dy}{dx} = \dfrac{1}{x^4-1}$.

8. $\dfrac{dy}{dx} = \dfrac{x+1}{x(1+x^2)}$.

9. $\dfrac{dy}{dx} = \dfrac{x}{x^4-x^2-2}$.

10. $\dfrac{dy}{dx} = \dfrac{1}{x(1+x)^2}$.

11. $\dfrac{dy}{dx} = \dfrac{3x+1}{x^4-1}$.

12. $\dfrac{dy}{dx} = \dfrac{x^4}{x^2-1}$.

13. Show that $\displaystyle\int \dfrac{\frac{du}{dx}}{u^2-a^2} = \dfrac{1}{2a} \log \dfrac{u-a}{u+a} + C$, u being a continuous function of x. This result is very useful.

LOGARITHMIC AND EXPONENTIAL FUNCTIONS 139

14. Integrate $\dfrac{dy}{dx} = \dfrac{1}{x^2 - 6x + 5}$. Writing $\dfrac{dy}{dx} = \dfrac{\frac{d}{dx}(x-3)}{(x-3)^2 - 4}$,

by Prob. 13, $y = \tfrac{1}{4}\log\dfrac{(x-3) - 2}{(x-3) + 2} + C = \tfrac{1}{4}\log\dfrac{x-5}{x-1} + C$.

15. Integrate $\dfrac{dy}{dx} = \dfrac{dx}{x^2 + 3x + 1}$.

ART. 49. — INTEGRATION BY PARTS

From $\dfrac{d}{dx}(u \cdot v) = u \cdot \dfrac{dv}{dx} + v \cdot \dfrac{du}{dx}$ is obtained by integration $\int u \cdot \dfrac{dv}{dx} = u \cdot v - \int v \cdot \dfrac{du}{dx}$, which is called the formula for integration by parts. The following examples will show the application of this formula.

EXAMPLE I. — Integrate $\dfrac{dy}{dx} = x \cdot \log x$.

Writing $u = \log x$, $\dfrac{dv}{dx} = x$, whence $\dfrac{du}{dx} = \dfrac{1}{x}$, $v = \tfrac{1}{2}x^2 + C_1$, the application of the formula gives

$$y = \int x \cdot \log x = (\tfrac{1}{2}x^2 + C_1) \cdot \log x - \int (\tfrac{1}{2}x^2 + C_1) \cdot \dfrac{1}{x}$$
$$= \tfrac{1}{2}x^2 \cdot \log x + C_1 \cdot \log x - \tfrac{1}{4}x^2 - C_1 \cdot \log x + C$$
$$= \tfrac{1}{2}x^2 \cdot \log x - \tfrac{1}{4}x^2 + C.$$

C_1, the constant of the integration $\int \dfrac{dv}{dx}$, always eliminates as in this example. It may therefore be neglected.

EXAMPLE II. — Integrate $\dfrac{dy}{dx} = x \cdot \sin x$.

Writing $u = x$, $\dfrac{dv}{dx} = \sin x$, whence $\dfrac{du}{dx} = 1$ and $v = -\cos x$, the application of the formula gives

$$y = \int x \cdot \sin x = -x \cdot \cos x + \int \cos x = -x \cdot \cos x + \sin x + C.$$

EXAMPLE III. — Integrate $\dfrac{dy}{dx} = x \cdot \sin^{-1} x$.

Writing $u = \sin^{-1} x$, $\dfrac{dv}{dx} = x$, whence $\dfrac{du}{dx} = \dfrac{1}{\sqrt{1-x^2}}$ and $v = \tfrac{1}{2} x^2$, the application of the formula gives

$$y = \int x \cdot \sin^{-1} x = \tfrac{1}{2} x^2 \cdot \sin^{-1} x - \tfrac{1}{2} \int \dfrac{x^2}{\sqrt{1-x^2}}.$$

Write $\dfrac{dy'}{dx} = \dfrac{x^2}{\sqrt{1-x^2}}$ and substitute $x = \sin \theta$. There results

$$\dfrac{dy'}{d\theta} = \dfrac{dy'}{dx} \cdot \dfrac{dx}{d\theta} = \sin^2 \theta = \tfrac{1}{2} - \tfrac{1}{2} \cos (2\theta),$$

and $\quad y' = \tfrac{1}{2}\theta - \tfrac{1}{4} \sin(2\theta) + C = \tfrac{1}{2}\theta - \tfrac{1}{2} \sin \theta \cdot \cos \theta + C.$

Substituting $\sin \theta = x$, $\cos \theta = \sqrt{1-x^2}$, $\theta = \sin^{-1} x$,

$$y' = \tfrac{1}{2} \sin^{-1} x - \tfrac{1}{2} x \cdot \sqrt{1-x^2} + C,$$

and $\quad y = \tfrac{1}{2} x^2 \cdot \sin^{-1} x + \tfrac{1}{4} \sin^{-1} x - \tfrac{1}{4} x \cdot \sqrt{1-x^2} + C.$

EXAMPLE IV. — Integrate $\dfrac{dy}{dx} = \sec^3 x$.

Writing $u = \sec x$, $\dfrac{dv}{dx} = \sec^2 x$, whence $\dfrac{du}{dx} = \sec x \cdot \tan x$ and $v = \tan x$, the application of the formula gives

$$y = \int \sec^3 x = \sec x \cdot \tan x - \int \sec x \cdot \tan^2 x$$

$$= \sec x \cdot \tan x - \int \sec^3 x + \int \dfrac{1}{\cos x}.$$

Hence $2 \int \sec^3 x = \sec x \cdot \tan x + \int \dfrac{1}{\cos x}$

$$= \sec x \cdot \tan x - \log \tan\left(\dfrac{\pi}{4} - \dfrac{x}{2}\right) + C$$

$$= \sec x \cdot \tan x + \log(\sec x - \tan x).$$

LOGARITHMIC AND EXPONENTIAL FUNCTIONS

EXAMPLE V. — Integrate $\dfrac{dy}{dx} = \sqrt{a^2 + x^2}$.

Writing $u = \sqrt{a^2 + x^2}$, $\dfrac{dv}{dx} = 1$, whence $\dfrac{du}{dx} = \dfrac{x}{\sqrt{a^2 + x^2}}$ and $v = x$, the application of the formula gives

$$(1) \quad y = x \cdot \sqrt{a^2 + x^2} - \int \dfrac{x^2}{\sqrt{a^2 + x^2}}.$$

Also $\dfrac{dy}{dx} = \dfrac{a^2 + x^2}{\sqrt{a^2 + x^2}} = \dfrac{a^2}{\sqrt{a^2 + x^2}} + \dfrac{x^2}{\sqrt{a^2 + x^2}}$,

hence $\quad (2) \quad y = a^2 \cdot \log(x + \sqrt{a^2 + x^2}) + \int \dfrac{x^2}{\sqrt{a^2 + x^2}}.$

Adding (1) and (2) and solving for y,

$$y = \tfrac{1}{2} x \cdot \sqrt{a^2 + x^2} + \dfrac{a^2}{2} \log(x + \sqrt{a^2 + x^2}) + C.$$

PROBLEMS

Integrate,

1. $\dfrac{dy}{dx} = x^2 \cdot \log x.$

2. $\dfrac{dy}{dx} = x^2 \cdot \cos x.$

3. $\dfrac{dy}{dx} = \tan^{-1} x.$

4. $\dfrac{dy}{dx} = x \cdot \tan^2 x.$

5. $\dfrac{dy}{dx} = \dfrac{\log^2 x}{x^{\frac{1}{2}}}.$

6. $\dfrac{dy}{dx} = \log x.$

7. $\dfrac{dy}{dx} = \dfrac{x \log x}{(a^2 + x^2)^{\frac{1}{2}}}.$

8. $\dfrac{dy}{dx} = \dfrac{x^2 \tan^{-1} x}{1 + x^2}.$

9. Find the area of the cycloid,
$$x = r \cdot \theta - r \cdot \sin \theta, \quad y = r - r \cdot \cos \theta.$$

10. Find the area of the cycloid,
$$x = r - r \cdot \cos \theta, \quad y = r \cdot \theta + r \cdot \sin \theta.$$

11. Find the volume of the solid generated by the revolution of the cycloid $x = r \cdot \theta - r \cdot \sin \theta$, $y = r - r \cdot \cos \theta$, about its base.

12. Find the volume of the solid generated by the revolution of the cycloid $x = r - r \cdot \cos \theta$, $y = r \cdot \theta + r \cdot \sin \theta$, about its axis.

13. Find the length of the parabola $y^2 = 2px$ from $x = 0$ to $x = x'$.

14. Find the length of the spiral $r = a \cdot \theta$ from $r = 0$ to $r = r'$.

15. Find the area bounded by the hyperbola $x^2 - y^2 = a^2$, the X-axis, and the ordinate to the point (x, y) of the hyperbola.

16. Show that the area bounded by the hyperbola $x^2 - y^2 = a^2$, the X-axis, and the line from $(0, 0)$ to the point (x, y) of the hyperbola is $\dfrac{a^2}{2} \log \dfrac{x+y}{a}$.

Art. 50. — Integration by Rationalization

A derivative $\dfrac{dy}{dx}$ containing the binomial $a + bx$ affected by fractional exponents may be transformed into a rational derivative $\dfrac{dy}{dz}$ by the substitution $a + bx = z^n$, where n is the least common multiple of the denominators of the fractional exponents.

Example I. — Integrate $\dfrac{dy}{dx} = \dfrac{x^{\frac{1}{2}}}{x^{\frac{1}{2}} - x^{\frac{1}{3}}}$.

Substituting $x = z^6$, $\dfrac{dx}{dz} = 6z^5$, $\dfrac{dy}{dz} = \dfrac{dy}{dx} \cdot \dfrac{dx}{dz} = \dfrac{6z^6}{z^3 - z^2} = \dfrac{6z^4}{z-1}$

$$= 6z^3 + 6z^2 + 6z + 6 + \dfrac{6}{z-1}.$$

Integrating, $y = \tfrac{3}{2} z^4 + 2 z^3 + 3 z^2 + 6 z + 6 \log(z-1) + C$;

whence, $\qquad y = \tfrac{3}{2} x^{\frac{2}{3}} + 2 x^{\frac{1}{2}} + 3 x^{\frac{1}{3}} + 6 x^{\frac{1}{6}} + 6 \log(x^{\frac{1}{6}} - 1) + C.$

LOGARITHMIC AND EXPONENTIAL FUNCTIONS 143

Example II. — Integrate $\dfrac{dy}{dx} = \dfrac{1}{(1+x)^{\frac{2}{3}} + (1+x)^{\frac{1}{2}}}$.

Substituting $1+x = z^2$, $\dfrac{dx}{dz} = 2z$, $\dfrac{dy}{dz} = \dfrac{dy}{dx} \cdot \dfrac{dx}{dz} = \dfrac{2z}{z^3+z} = \dfrac{2}{z^2+1}$.

Integrating, $y = 2\tan^{-1}z + C$; whence, $y = 2\tan^{-1}(1+x)^{\frac{1}{2}} + C$.

A derivative $\dfrac{dy}{dx}$ containing only the surd $\sqrt{a+bx+x^2}$ may be transformed into a rational derivative $\dfrac{dy}{dz}$ by the substitution $\sqrt{a+bx+x^2} = z - x$.

A derivative $\dfrac{dy}{dx}$ containing only the surd $\sqrt{a+bx-x^2}$ may be transformed into a rational derivative $\dfrac{dy}{dz}$ by the substitution $\sqrt{a+bx-x^2} = \sqrt{(x-r_1)(r_2-x)} = (x-r_1) \cdot z$.

Example III. — Integrate $\dfrac{dy}{dx} = \dfrac{\sqrt{2x+x^2}}{x^2}$.

Writing

$$\sqrt{2x+x^2} = z - x, \quad x = \dfrac{z^2}{2z+2}, \quad \dfrac{dx}{dz} = \dfrac{2z^2+4z}{2z+2}, \quad z - x = \dfrac{z^2+2z}{2z+2}.$$

Hence,

$$\dfrac{dy}{dz} = \dfrac{dy}{dx} \cdot \dfrac{dx}{dz} = \dfrac{z^2+4z+4}{z^2(z+1)} = \dfrac{z^2}{z^2(z+1)} + \dfrac{4(z+1)}{z^2(z+1)} = \dfrac{1}{z+1} + \dfrac{4}{z^2}.$$

Integrating,

$$y = \log(z+1) - \dfrac{4}{z} + C = \log(x+1+\sqrt{2x+x^2}) - \dfrac{4}{x+\sqrt{2x+x^2}} + C.$$

Example IV. — Integrate $\dfrac{dy}{dx} = \dfrac{1}{x\sqrt{2+x-x^2}}$.

Writing $\sqrt{2+x-x^2} \equiv \sqrt{(2-x)(1+x)} = (2-x) \cdot z$,

$$x = \dfrac{2z^2-1}{z^2+1}, \quad \dfrac{dx}{dz} = \dfrac{6z}{(z^2+1)^2}, \quad \text{and} \quad \dfrac{dy}{dz} = \dfrac{2}{2z^2-1}.$$

By partial fractions, $\dfrac{dy}{dz} = \dfrac{1}{z\sqrt{2}-1} - \dfrac{1}{z\sqrt{2}+1}$.

Integrating, $y = \dfrac{1}{\sqrt{2}} \log(\sqrt{2}\cdot z - 1) - \dfrac{1}{\sqrt{2}} \log(\sqrt{2}\cdot z + 1) + C.$

PROBLEMS

Integrate,

1. $\dfrac{dy}{dx} = x^2 \cdot (1+x)^{\frac{1}{2}}.$

2. $\dfrac{dy}{dx} = \dfrac{x^3}{\sqrt{x-1}}.$

3. $\dfrac{dy}{dx} = \dfrac{x^3}{(1+4x)^{\frac{3}{2}}}.$

4. $\dfrac{dy}{dx} = \dfrac{1}{x\cdot\sqrt{1+x}}.$

5. $\dfrac{dy}{dx} = \dfrac{1}{x+\sqrt{x-1}}.$

6. $\dfrac{dy}{dx} = \dfrac{x^{\frac{1}{2}}}{x^{\frac{3}{4}}+1}.$

7. $\dfrac{dy}{dx} = \dfrac{1}{x^{\frac{1}{5}}+x^{\frac{1}{3}}}.$

8. $\dfrac{dy}{dx} = \dfrac{\sqrt{6x-x^2}}{x}.$

9. $\dfrac{dy}{dx} = \dfrac{x}{(2+3x-2x^2)^{\frac{3}{2}}}.$

10. $\dfrac{dy}{dx} = \dfrac{1}{x\sqrt{x^2+2x-1}}.$

ART. 51. — EVALUATION OF FORMS 1^∞, ∞^0, 0^0

If $y = f(x)^{\phi(x)} = 1^\infty$ when $x = a$,

$\log_e y = \phi(x) \cdot \log_e f(x) = \infty \cdot 0$ when $x = a$.

If $y = f(x)^{\phi(x)} = \infty^0$ when $x = a$,

$\log_e y = \phi(x) \cdot \log_e f(x) = 0 \cdot \infty$ when $x = a$.

If $y = f(x)^{\phi(x)} = 0^0$ when $x = a$,

$\log_e y = \phi(x) \cdot \log_e f(x) = 0 \cdot \infty$ when $x = a$.

The true value of $\log_e y$ is found by evaluating the form $\infty \cdot 0$, and the true value of y becomes known.

LOGARITHMIC AND EXPONENTIAL FUNCTIONS

EXAMPLE. — Find the true value of $\left(\dfrac{a}{x}+1\right)^x$ when $x=\infty$.
Here,
$$y=\left(\dfrac{a}{x}+1\right)^x=1^\infty \text{ and } \log_e y = x\cdot\log_e\left(\dfrac{a}{x}+1\right)=\infty\cdot 0 \text{ when } x=\infty.$$

Hence, $\log_e y = \dfrac{\log_e\left(\dfrac{a}{x}+1\right)}{\dfrac{1}{x}} = \dfrac{\dfrac{a}{x}+1}{-\dfrac{1}{x^2}} \cdot \left(-\dfrac{a}{x^2}\right) = \dfrac{a}{\dfrac{a}{x}+1} = a$ when $x=\infty$.

Since $\log_e y = a$, $y = e^a$.

PROBLEMS

Find the true value of,

1. $\left(1+\dfrac{1}{x^2}\right)^x$ when $x=\infty$.
2. $x^{\frac{1}{1-x}}$ when $x=1$.
3. $(\sin x)^{\sin x}$ when $x=0$.
4. $(\sin x)^{\tan x}$ when $x=\dfrac{\pi}{2}$.
5. $(1+nx)^{\frac{1}{x}}$ when $x=0$.
6. $\{\cos(ax)\}^{\operatorname{cosec}^2(cx)}$ when $x=0$.
7. $\left(\dfrac{1}{x}\right)^x$ when $x=0$.
8. $(\cos 2x)^{\frac{1}{x^2}}$ when $x=0$.
9. $(\log x)^{x-1}$ when $x=1$.

ART. 52. — DIFFERENTIATION OF EXPONENTIAL FUNCTIONS

The logarithm of the exponential function a^u is
$$\log_e a^u = u\cdot\log_e a.$$
Differentiating,
$$\dfrac{1}{a^u}\cdot\dfrac{d}{dx}a^u = \log_e a\cdot\dfrac{du}{dx}; \text{ whence, } \dfrac{d}{dx}a^u = a^u\cdot\log_e a\cdot\dfrac{du}{dx}.$$

That is, the derivative of an exponential function with a constant base is the product of the exponential function, the logarithm of the base, and the derivative of the exponent.

If $a = e$, $\dfrac{d}{dx} e^u = e^u \cdot \dfrac{du}{dx}$.

EXAMPLE. — Differentiate, $y = a^{x^2}$.

Here $\dfrac{dy}{dx} = a^{x^2} \cdot \log_e a \cdot \dfrac{d}{dx} x^2 = 2x \cdot a^{x^2} \cdot \log_e a$.

PROBLEMS

Differentiate,

1. $y = a^{\frac{1}{x}}$.   2. $y = a^{\frac{1}{1-x}}$.   3. $y = a^{\sin x}$.   4. $y = a^{\tan^{-1} x}$.

5. $y = e^{ax}$.   6. $y = e^{(1+x^2)}$.   7. $y = e^{\sin^{-1} x}$.

8. Show that $\dfrac{d^u}{dx^u} e^{ax} = a^u \cdot e^{ax}$.

Differentiate, 9. $y = x^x$. Here $\log y = x \cdot \log x$, and by differentiation $\dfrac{1}{y} \cdot \dfrac{dy}{dx} = 1 + \log x$. Hence, $\dfrac{dy}{dx} = x^x \cdot (1 + \log x)$.

10. $y = e^{x^x}$.   11. $y = e^{x^x}$.   12. $y = x^{x^x}$.

Find the true value of,

13. $(e^x + 1)^{\frac{1}{x}}$ when $x = 0$.

14. $\dfrac{a^x - b^x}{x}$ when $x = 0$.

15. $\dfrac{e^{nx} - e^{na}}{(x-a)}$ when $x = a$.

16. $\dfrac{e^x - e^{-x} - 2x}{x - \sin x}$ when $x = 0$.

17. $\dfrac{x^n}{e^x}$ when $x = \infty$.

18. $e^{\frac{1}{x}} \cdot \sin x$ when $x = 0$.

19. $\dfrac{e^x - e^{-x}}{\sin x}$ when $x = 0$.

20. Show that $x^{\frac{1}{x}}$ is a minimum when $x = e$.

21. Find the least value of $ae^{kx} + be^{-kx}$.

ART. 53. — INTEGRATION OF EXPONENTIAL FUNCTIONS

The results of the preceding article may be written

$$\int a^u \cdot \frac{du}{dx} = \frac{a^u}{\log_e a} + C, \quad \int e^u \cdot \frac{du}{dx} = e^u + C.$$

EXAMPLE. — Integrate $\frac{dy}{dx} = e^{3x-2}$.

Writing $u = 3x - 2$, $\frac{du}{dx} = 3$ and $\frac{dy}{dx} = \frac{1}{3} e^{3x-2} \cdot \frac{d}{dx}(3x - 2)$.
Integrating, $y = \frac{1}{3} \cdot e^{3x-2} + C$.

PROBLEMS

Integrate,

1. $\frac{dy}{dx} = e^{\frac{x}{3}}$.
2. $\frac{dy}{dx} = e^{2x}$.
3. $\frac{dy}{dx} = e^{nx}$.
4. $\frac{dy}{dx} = x \cdot e^{x^2}$.
5. $\frac{dy}{dx} = x^2 \cdot e^{x^3}$.
6. $\frac{dy}{dx} = x^{-2} \cdot e^{\frac{1}{x}}$.
7. $\frac{dy}{dx} = a^{3x}$.
8. $\frac{dy}{dx} = a^{mx}$.
9. $\frac{dy}{dx} = a^{2x+5}$.
10. $\frac{dy}{dx} = a^{3-5x}$.
11. $\frac{dy}{dx} = e^x \cdot \sin x$.

Applying the formula for integration by parts by writing

$u = \sin x$, $\frac{dv}{dx} = e^x$, $y = \int e^x \cdot \sin x = e^x \cdot \sin x - \int e^x \cdot \cos x$.

Applying the formula to $\int e^x \cdot \cos x$ by writing $u = \cos x$,
$\frac{dv}{dx} = e^x$,
$$\int e^x \cdot \cos x = -e^x \cdot \cos x + \int e^x \cdot \sin x,$$
Hence
$$\int e^x \cdot \sin x = e^x \cdot \sin x + e^x \cdot \cos x - \int e^x \cdot \sin x,$$
and
$$y = \int e^x \cdot \sin x = \frac{1}{2} e^x \cdot \sin x + \frac{1}{2} e^x \cdot \cos x + C.$$

12. $\dfrac{dy}{dx} = e^x \cdot \cos x$.   13. $\dfrac{dy}{dx} = e^x \cdot x^2$.   14. $\dfrac{dy}{dx} = a^x \cdot x^3$.

15. $\dfrac{dy}{dx} = e^x \cdot x^{-2}$.   16. $\dfrac{dy}{dx} = e^{-x} \cdot x^3$.

Art. 54. — The Hyperbolic Functions

The functions $\frac{1}{2}(e^x + e^{-x})$ and $\frac{1}{2}(e^x - e^{-x})$ are called the hyperbolic cosine of x and the hyperbolic sine of x respectively, and are denoted by the symbols $\cosh x$ and $\sinh x$. Hence by definition $\cosh x = \frac{1}{2}(e^x + e^{-x})$, $\sinh x = \frac{1}{2}(e^x - e^{-x})$. It follows at once that $(\cosh x)^2 - (\sinh x)^2 = 1$.

The inverse hyperbolic functions are denoted by $\cosh^{-1} x$ and $\sinh^{-1} x$, so that $y = \cosh^{-1} x$ and $y = \sinh^{-1} x$ are equivalent to $x = \cosh y$ and $x = \sinh y$ respectively.

The hyperbolic functions have been calculated and tabulated.

Example I. — Find the derivative of $y = \sinh x$.

Differentiating,

$$\frac{dy}{dx} = \frac{d}{dx} \sinh x = \frac{d}{dx} \tfrac{1}{2}(e^x - e^{-x}) = \tfrac{1}{2}(e^x + e^{-x}) = \cosh x.$$

Example II. — Differentiate $y = \sinh^{-1} x$.

Writing $x = \sinh y$ and differentiating,

$$1 = \cosh y \cdot \frac{dy}{dx}, \text{ whence } \frac{dy}{dx} = \frac{1}{\cosh y} = \frac{1}{\sqrt{1 + \sinh^2 y}} = \frac{1}{\sqrt{1 + x^2}}.$$

Hence $\dfrac{d}{dx} \sinh^{-1} x = \dfrac{1}{\sqrt{1 + x^2}}$.

PROBLEMS

1. Show that $\dfrac{d}{dx} \cosh x = \sinh x$.

2. Show that $\dfrac{d}{dx} \sinh^{-1} \dfrac{x}{a} = \dfrac{1}{\sqrt{a^2 + x^2}}$.

3. Show that $\dfrac{d}{dx}\cosh^{-1}\dfrac{x}{a} = \dfrac{1}{\sqrt{x^2 - a^2}}$.

4. Find the minimum value of $\cosh x$.

5. An inextensible, perfectly flexible, homogeneous string fastened at two points in the same horizontal is acted on by gravity only. It is shown in mechanics that for any point (x, y) in the position of equilibrium $\dfrac{dy}{dx} = \dfrac{s}{c}$, where s is the length of string from the lowest point to (x, y), and c is the length of string whose weight equals the tension at the lowest point. Find equation of position of equilibrium.

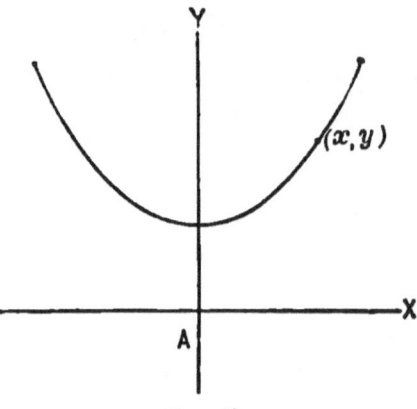

Fig. 48.

The x-derivative of $\dfrac{dy}{dx} = \dfrac{s}{c}$

is $\dfrac{d^2y}{dx^2} = \dfrac{\dfrac{ds}{dx}}{c} = \dfrac{1}{c}\sqrt{1 + \dfrac{dy^2}{dx^2}}$, whence $\dfrac{\dfrac{d}{dx}\dfrac{dy}{dx}}{\sqrt{1 + \dfrac{dy^2}{dx^2}}} = \dfrac{1}{c}$.

Integrating, $\sinh^{-1}\dfrac{dy}{dx} = \dfrac{x}{c} + C_1$. When $x = 0$, $\dfrac{dy}{dx} = 0$, hence $C_1 = 0$, and $\dfrac{dy}{dx} = \sinh\dfrac{x}{c}$. Integrating again, $y = c\cdot\cosh\dfrac{x}{c} + C_2$. In the figure, for $x = 0$, $y = c$, hence $C_2 = 0$, and finally $y = c\cdot\cosh\dfrac{x}{c}$, the equation of the catenary. Using exponential functions,
$$y = \dfrac{c}{2}(e^{\frac{x}{c}} + e^{-\frac{x}{c}}).$$

6. Find the length of the catenary $y = c\cdot\cosh\dfrac{x}{c}$ from $x = 0$ to $x = x'$.

7. Find the radius of curvature of the catenary.

Art. 55.—The Definite Integral $\int_{-\infty}^{+\infty} e^{-x^2} \cdot dx$

The definite integral $\int_{-\infty}^{+\infty} e^{-x^2} \cdot dx$ is very important in mathematical physics and the theory of probability. Its value may be determined by the following analysis, due to Poisson.

Denoting the value of the integral by A, $A = \int_{-\infty}^{+\infty} e^{-x^2} \cdot dx$, $A = \int_{-\infty}^{+\infty} e^{-y^2} \cdot dy$, where x and y are assumed to be independent variables. Hence,

$$A^2 = \int_{-\infty}^{+\infty} e^{-x^2} \cdot dx \cdot \int_{-\infty}^{+\infty} e^{-y^2} \cdot dy = \int_{y=-\infty}^{y=+\infty} \int_{x=-\infty}^{x=+\infty} e^{-(x^2+y^2)} \cdot dx\, dy.$$

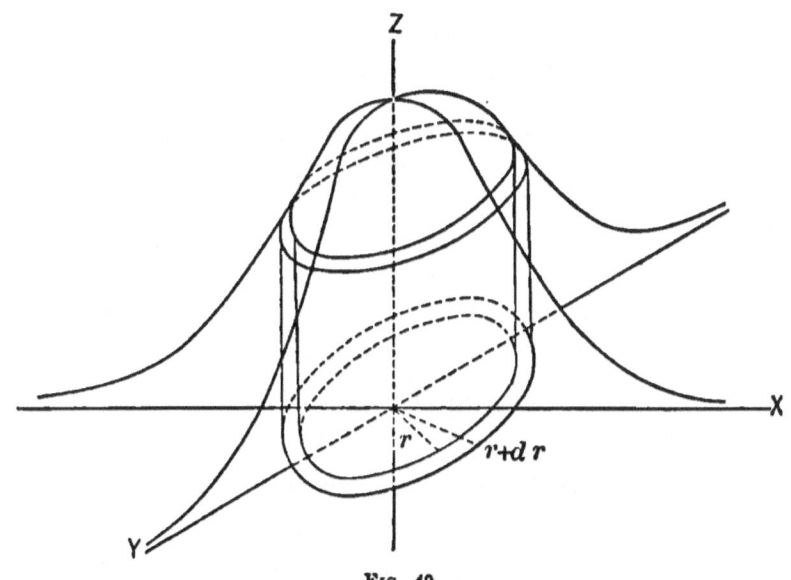

Fig. 49.

Now, $z = e^{-(x^2+y^2)}$ represents the surface of revolution whose generatrix in the ZX-plane is $z = e^{-x^2}$ and axis of revolution the Z-axis. Hence, $\int_{y=-\infty}^{y=+\infty} \int_{x=-\infty}^{x=+\infty} e^{-(x^2+y^2)} \cdot dx\, dy$ is the volume bounded by this surface of revolution and the XY-plane.

An element of the volume of this solid is the cylindric shell of thickness dr, radius r, altitude $e^{-(x^2+y^2)} = e^{-r^2}$. Hence the volume of the solid is $2\pi \int_0^{\infty} e^{-r^2} \cdot r\, dr = \pi$. Therefore, $A^2 = \pi$ and $A = \int_{-\infty}^{+\infty} e^{-x^2} \cdot dx = \sqrt{\pi}$.

To determine $\int_{-\infty}^{+\infty} e^{-(x^2+2ax)} \cdot dx$, write $x^2 + 2ax = (x+a)^2 - a^2$; whence, $\int_{-\infty}^{+\infty} e^{-(x^2+2ax)} \cdot dx = e^{a^2} \int_{-\infty}^{+\infty} e^{-(x+a)^2} d(x+a) = \sqrt{\pi} \cdot e^{a^2}$.

ART. 56. — DIFFERENTIATION OF A DEFINITE INTEGRAL

Denoting the definite integral $\int_0^3 a^2 \cdot x^2 \cdot dx$ by A, $A = 3\,a^2$ and $\dfrac{dA}{da} = 6\,a$. Differentiating $a^2 \cdot x^2 \cdot dx$ with respect to a, and integrating the result with respect to x between the limits $x = 0$, $x = 3$, $\int_0^3 \dfrac{d}{da}(a^2 x^2\, dx) = \int_0^3 2\,ax^2\, dx = 6\,a$. Hence for this special definite integral, $\dfrac{d}{da}\int_0^3 a^2 \cdot x^2 \cdot dx = \int_0^3 \dfrac{d}{da}(a^2 x^2\, dx)$. That is, to differentiate the definite integral with respect to a parameter, differentiate the function under the sign of integration with respect to the parameter.

To prove this proposition in general, consider the definite integral $A = \int_c^b f(x, a)\, dx$. If $f(x, a)$ is a continuous function of a, $\int_c^b f(x, a)\, dx$ is also a continuous function of a. For, denoting by ΔA and Δa the corresponding changes of A and a, $\Delta A = \int_c^b \{f(x, a + \Delta a) - f(x, a)\}\, dx = \epsilon \int_c^b dx = \epsilon(b-c)$, where ϵ is a quantity which approaches zero when Δa approaches zero. Hence A is a continuous function of a. The derivative of A with respect to a is

$$\frac{dA}{da} = \text{limit} \int_c^b \frac{\{f(x, a+\Delta a) - f(x, a)\}}{\Delta a} dx = \int_c^b \frac{d}{da} f(x, a) \cdot dx,$$

which proves the general proposition.

By this proposition the values of definite integrals of a general form may be found from known special definite integrals.

EXAMPLE I. — The definite integral $\int_0^\infty e^{-ax} \cdot dx = \frac{1}{a}$.

Differentiating with respect to a,

$$\int_0^\infty x \cdot e^{-ax} \cdot dx = \frac{1}{a^2}, \qquad \int_0^\infty x^2 \cdot e^{-ax} \cdot dx = \frac{1 \cdot 2}{a^3},$$

$$\int_0^\infty x^3 \cdot e^{-ax} \cdot dx = \frac{1 \cdot 2 \cdot 3}{a^4}, \cdots, \int_0^\infty x^n \cdot e^{-ax} \cdot dx = \frac{1 \cdot 2 \cdot 3, \cdots n}{a^{n+1}}.$$

EXAMPLE II. — The definite integral $\int_0^\infty \frac{dx}{x^2 + a^2} = \frac{\pi}{2} \cdot \frac{1}{a}$.

Differentiating with respect to a,

$$\int_0^\infty \frac{dx}{(x^2+a^2)^2} = \frac{\pi}{2} \cdot \frac{1}{a} \cdot \frac{1}{2}, \qquad \int_0^\infty \frac{dx}{(x^2+a^2)^3} = \frac{\pi}{2} \cdot \frac{1}{a} \cdot \frac{1}{2} \cdot \frac{3}{4}, \cdots,$$

$$\int_0^\infty \frac{dx}{(x^2+a^2)^{n+1}} = \frac{\pi}{2} \cdot \frac{1}{a} \cdot \frac{1}{2} \cdot \frac{3}{4} \cdots \frac{2n-1}{2n}.$$

ART. 57. — MEAN VALUE

Let it be required to determine the mean or average value of the continuous function $f(x)$ from $x = a$ to $x = b$. This is equivalent to finding the mean ordinate of the curve $y = f(x)$ from $x = a$ to $x = b$. Divide the portion of the X-axis from $x = a$ to $x = b$ into n equal parts Δx, and draw an ordinate of $y = f(x)$ at the end of each Δx nearest the origin. Denoting the sum of these n ordinates by Σy, their mean value is $\frac{\Sigma y}{n} = \frac{\Sigma y \cdot \Delta x}{n \cdot \Delta x}$. This is true for all values of n. When n is indefinitely increased Δx becomes dx, Σy becomes the sum of all the ordinates of $y = f(x)$ from $x = a$ to $x = b$, and the mean value of all these ordinates becomes $\dfrac{\int_a^b y\,dx}{n \cdot dx} = \dfrac{\int_a^b y\,dx}{b - a}$, since $n \cdot \Delta x = b - a$ for all values of n.

EXAMPLE. — Find the mean ordinate of the sine curve $y = \sin x$ from $x = 0$ to $x = \pi$.

Here mean ordinate $= \dfrac{\int_0^\pi \sin x \cdot dx}{\pi} = \dfrac{2}{\pi}$.

PROBLEMS

1. Find mean ordinate of circle $x^2 + y^2 = r^2$ from $x = +r$ to $x = -r$.

2. Find mean ordinate of ellipse $\dfrac{x^2}{a^2} + \dfrac{y^2}{b^2} = 1$ from $x = +a$ to $x = -a$.

3. Find mean value of $\sin^2 x$ from $x = 0$ to $x = \pi$.

CHAPTER IX

CENTER OF MASS AND MOMENT OF INERTIA

Art. 58. — Center of Mass

If the mass of a body is divided into infinitesimal elements of mass dm and the coordinates of dm are x, y, z, the center of mass of the body is the point $(\bar{x}, \bar{y}, \bar{z})$ so situated that $\bar{x}$ multiplied by the entire mass of the body is equal to the sum of the products of each element of mass dm by the distance of this element of mass from the YZ-plane, with like definitions for $\bar{y}$ and $\bar{z}$. Hence the coordinates of the center of mass of a body are $\bar{x} = \dfrac{\int x\,dm}{\int dm}$, $\bar{y} = \dfrac{\int y\,dm}{\int dm}$, $\bar{z} = \dfrac{\int z\,dm}{\int dm}$, the integration extending over the entire mass of the body.*

Representing the magnitude of the element of mass dm by dM, its density by D, $dm = D \cdot dM$, and the coordinates of the center of mass become

---

* In this chapter differentials are used directly. Lagrange says in the preface to his Mecanique Analytique (1811), "When we have properly conceived the spirit of the infinitesimal method, and are convinced of the exactness of its results by the geometrical method of prime and ultimate ratios, or by the analytical method of derived functions, we may employ infinitely small quantities as a sure and valuable means of abridging and simplifying our demonstrations."

CENTER OF MASS AND MOMENT OF INERTIA 155

$$\bar{x} = \frac{\int x D dM}{\int D dM}, \quad \bar{y} = \frac{\int y D dM}{\int D dM}, \quad \bar{z} = \frac{\int z D dM}{\int D dM},$$

the integration extending over the entire magnitude of the body.

If the body is homogeneous, D is constant, and

$$\bar{x} = \frac{\int x dM}{\int dM}, \quad \bar{y} = \frac{\int y dM}{\int dM}, \quad \bar{z} = \frac{\int z dM}{\int dM}.$$

The center of mass now becomes the center of figure. If the YZ-plane is a plane of symmetry of the body, to every term $+ x dM$ of $\int x dM$ there corresponds a term $- x dM$. Hence $\bar{x} = 0$; that is, the center of mass lies in the plane of symmetry of the body. In like manner it is shown that the center of mass lies in the axis of symmetry of the body and at the center of symmetry of the body. Unless otherwise specified, the density is assumed uniform; that is, the body is homogeneous.

In mechanics it is proved the center of gravity of a body coincides with its center of mass.

ART. 59. — CENTER OF MASS OF LINES

EXAMPLE I. — Find the center of mass of a straight line of length l, whose density varies as the distance from one end.

Denoting by dx an element of the length of the line, by x the distance of dx from the end A, by k the density of the line at unit's distance from A, whence

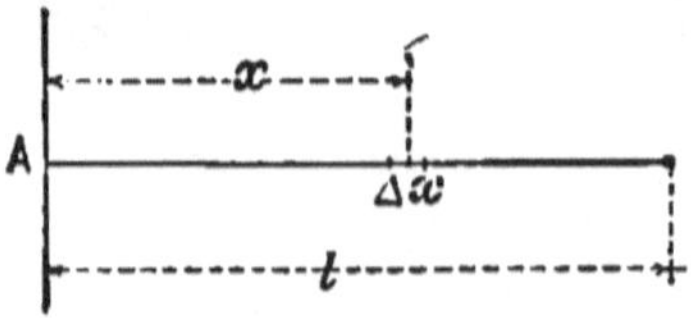

FIG. 50.

$$\frac{D}{k} = \frac{x}{1}, \quad \bar{x} = \frac{\int_0^l kx \cdot x \, dx}{\int_0^l kx \cdot dx} = \tfrac{2}{3} l.$$

EXAMPLE II. — Find the center of mass of the arc of a circle.

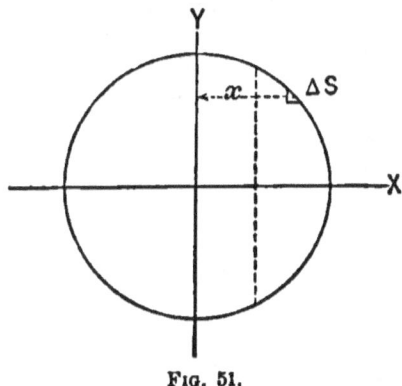

FIG. 51.

Take as X-axis the axis of symmetry of the arc. The element of magnitude is

$$ds = \left(1 + \frac{dx^2}{dy^2}\right)^{\frac{1}{2}} dy.$$

From the equation of the circle $x^2 + y^2 = R^2$,

$$\frac{dx}{dy} = -\frac{y}{x}.$$

Hence $ds = \dfrac{R}{x} dy$, and $\bar{x} = \dfrac{R \int_{-\frac{1}{2}\text{chord}}^{+\frac{1}{2}\text{chord}} dy}{\text{arc}} = \dfrac{\text{radius} \times \text{chord}}{\text{arc}}.$

PROBLEMS

1. Find the center of mass of the straight line of length l, whose density varies as the square of the distance from one end.

2. Find the center of mass of the entire cycloidal arc.

3. Find the center of mass of the length of one loop of the lemniscate $r^2 = a^2 \cos(2\theta)$, calling the length of the loop $2l$. Here $dM = ds = \left(r^2 + \dfrac{dr^2}{d\theta^2}\right)^{\frac{1}{2}} \cdot d\theta$ and $x = r \cdot \cos \theta$.

4. Find the center of mass of the quarter of the curve $x^{\frac{2}{3}} + y^{\frac{2}{3}} = a^{\frac{2}{3}}$ included by the coordinate axes.

5. Find the center of mass of the helix $x = a \cdot \sin \phi$, $y = a \cdot \cos \phi$, $z = c \cdot \phi$, from $\phi = 0$ to $\phi = \phi'$.

Here $dM = ds = \left(\dfrac{dx^2}{d\phi^2} + \dfrac{dy^2}{d\phi^2} + \dfrac{dz^2}{d\phi^2}\right)^{\frac{1}{2}} \cdot d\phi$.

6. Find the center of mass of a triangle.

Break up the triangle into infinitesimal strips by lines parallel to the base at intervals dx measured on the median to the base, and call the distance from the vertex to any strip measured on the median x. The median is an axis of symmetry of the triangle, and the center of mass must lie in the median. Concentrate the magnitude of each strip on the median, and the median becomes a line whose density varies as the distance from the vertex. Hence, k representing the density at unit's distance from the vertex, $x = \dfrac{\displaystyle\int_0^l kx \cdot x\, dx}{\displaystyle\int_0^l kx \cdot dx} = \tfrac{2}{3} l.$

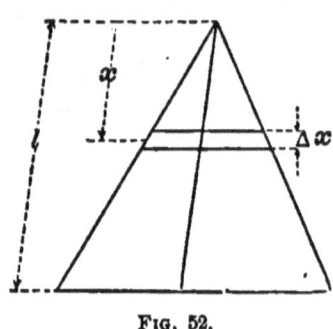

Fig. 52.

Art. 60. — Center of Mass of Surfaces

Example I. — Find the center of mass of the surface bounded by the parabola $y^2 = 2px$, the Y-axis, and the abscissa to the point (x_0, y_0) of the parabola.

The surface is broken up into strips of breadth dy by lines parallel to the

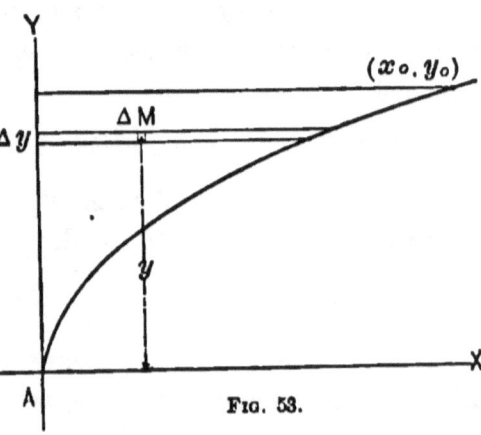

Fig. 53.

X-axis; each strip is broken up into elements of area dA by lines parallel to the Y-axis at intervals dx. Hence $dM = dA = dx\,dy$ and

$$\bar{x} = \frac{\int_{y=0}^{y=y_0}\int_{x=0}^{x=\frac{y^2}{2p}} x\,dx\,dy}{\int_{y=0}^{y=y_0}\int_{x=0}^{x=\frac{y^2}{2p}} dx\,dy} = \tfrac{3}{10} x_0,$$

$$\bar{y} = \frac{\int_{y=0}^{y=y_0}\int_{x=0}^{x=\frac{y^2}{2p}} y\,dx\,dy}{\int_{y=0}^{y=y_0}\int_{x=0}^{x=\frac{y^2}{2p}} dx\,dy} = \tfrac{3}{4} y_0,$$

since the object of the x-integration is to sum up the products $y\,dx\,dy$ for the elements of area $dx\,dy$ forming the strip, and the object of the y-integration is to sum up the strips forming the given surface.

EXAMPLE II. — Find the center of mass of the surface of the cycloid

$$x = r - r\cdot\cos\theta,\ \ y = r\cdot\theta + r\cdot\sin\theta.$$

The center of mass lies on the X-axis and

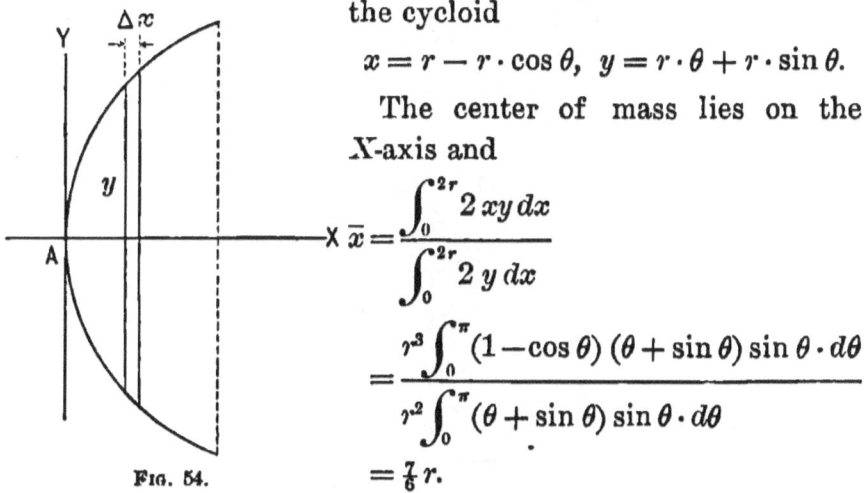

Fig. 54.

$$\bar{x} = \frac{\int_0^{2r} 2\,xy\,dx}{\int_0^{2r} 2\,y\,dx}$$

$$= \frac{r^3\int_0^{\pi}(1-\cos\theta)(\theta+\sin\theta)\sin\theta\cdot d\theta}{r^2\int_0^{\pi}(\theta+\sin\theta)\sin\theta\cdot d\theta}$$

$$= \tfrac{7}{6}r.$$

EXAMPLE III. — Find the center of mass of the circular sector whose angle is $2\theta_0$ and whose density varies as the square of the distance from the center.

Here it is advisable to use polar coordinates. Drawing radii at angular intervals $d\theta$, the sector is broken up into infinitesimal sectors. Call distances measured out from the center ρ and draw circles concentric at A at intervals $d\rho$. Each infinitesimal sector is divided into elements of area

$$dA = \rho \, d\theta \cdot d\rho$$

and $\quad x = \rho \cdot \cos \theta.$

Denoting the density at unit's distance from center by k,

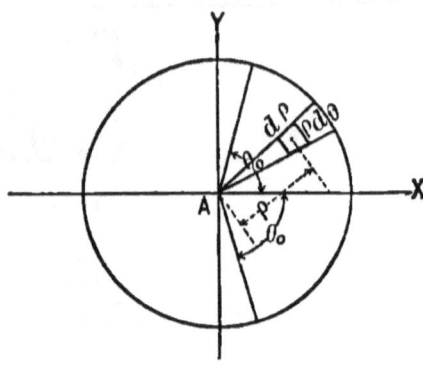

Fig. 55.

$$\frac{D}{k} = \frac{\rho^2}{1} \text{ and } D = k \cdot \rho^2. \text{ Hence}$$

$$\bar{x} = \frac{\int_{-\theta_0}^{+\theta_0}\int_0^R k \cdot \rho^2 \cdot \rho \cos\theta \cdot \rho \cdot d\theta d\rho}{\int_{-\theta_0}^{+\theta_0}\int_0^R k \cdot \rho^2 \cdot \rho \cdot d\theta d\rho} = \frac{4}{5}\frac{R \cdot \sin\theta}{\theta}.$$

EXAMPLE IV. — Find the center of mass of the eighth of

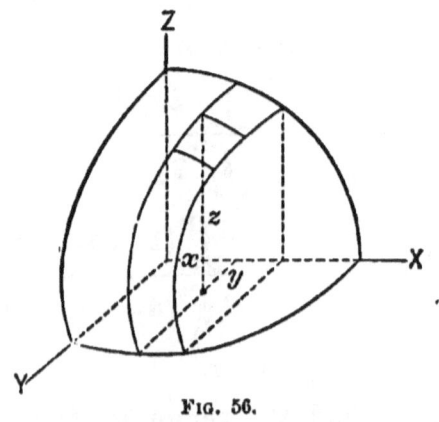

Fig. 56.

the surface of the sphere $x^2 + y^2 + z^2 = R^2$ bounded by the coordinate planes.

Here $dA = \left(1 + \dfrac{\partial z^2}{\partial x^2} + \dfrac{\partial z^2}{\partial y^2}\right)^{\frac{1}{2}} dx\, dy = \dfrac{R}{z} dx\, dy,$

hence $\bar{x} = \dfrac{\int_0^R \int_0^{(R^2-x^2)^{\frac{1}{2}}} \dfrac{Rx}{z} dx\, dy}{\frac{1}{2}\pi R^2} = \dfrac{2}{\pi R} \int_0^R \int_0^{(R^2-x^2)^{\frac{1}{2}}} \dfrac{x\, dx\, dy}{\sqrt{R^2 - x^2 - y^2}}$

$= \dfrac{2}{\pi R} \int_0^R x\, dx \left(\sin^{-1} \dfrac{y}{(R^2 - x^2)^{\frac{1}{2}}}\right)_0^{(R^2-x^2)^{\frac{1}{2}}}$

$= \dfrac{2\pi}{2\pi R} \int_0^R x\, dx = \tfrac{1}{2} R.$

By the nature of the problem $\bar{x} = \bar{y} = \bar{z}$.

PROBLEMS

1. Find the center of mass of the circular sector.
2. Find the center of mass of the quarter of the circle $x^2 + y^2 = R^2$ included by the coordinate axes.
3. Find the center of mass of the quarter of the ellipse $\dfrac{x^2}{a^2} + \dfrac{y^2}{b^2} = 1$ included by the coordinate axes.
4. Find the center of mass of the surface bounded by the parabola $y^2 = 2px$ and the double ordinate to the point (x, y).
5. Find the center of mass of the circular segment bounded by $y^2 = 2Rx - x^2$ and the double ordinate through (x, y).
6. Find the center of mass of the surface bounded by the circle $y^2 = 2Rx - x^2$, the Y-axis, and the abscissa to the point (x, y). This surface is called the circular spandrel.
7. Find the center of mass of the part of a circular annulus bounded by the circles $r = R$, $r = R'$, and the radii vectores $\theta = -\theta_0$, $\theta = +\theta_0$.
8. Find the center of mass of the surface bounded by $x^{\frac{2}{3}} + y^{\frac{2}{3}} = a^{\frac{2}{3}}$ and the positive coordinate axes.
9. Find the center of mass of the area of one loop of the lemniscate.
10. Find the center of mass of a zone.

Art. 61. — Center of Mass of Solids

Example I. — Find the center of mass of the half of the ellipsoid $\frac{x^2}{a^2}+\frac{y^2}{b^2}+\frac{z^2}{c^2}=1$ lying to the right of the ZY-plane.

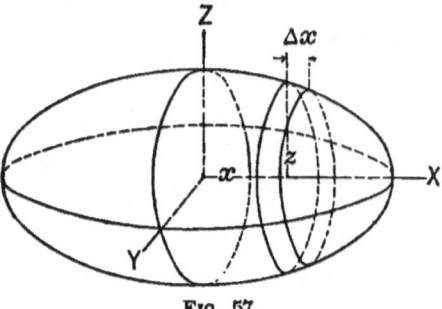

Fig. 57.

The X-axis is an axis of symmetry, and

$$dM = \pi rs \cdot rt \cdot dx$$
$$= \frac{\pi bc}{a^2}(a^2 - x^2)dx.$$

Hence $\quad \bar{x} = \dfrac{\pi \dfrac{bc}{a^2}\int_0^a (a^2-x^2)dx}{\tfrac{2}{3}\pi abc} = \tfrac{3}{8}a.$

Example II. — Find the center of mass of the eighth of the sphere $x^2 + y^2 + z^2 = R^2$ included by the coordinate planes, the density varying as the square of the distance from the center.

Here it is advisable to use polar coordinates. Passing planes through the Z-axis at angular intervals $d\phi$, the solid is divided into spherical wedges. Passing conical surfaces with vertex at O, and whose elements make angles with the XY-plane increasing by $d\theta$, each wedge is divided into pyramids. Passing spherical surfaces concentric at O, and whose radii increase by $d\rho$, each pyramid is divided into elementary parallelopipeds, whose dimensions are $d\rho$, $\rho \cos\theta\, d\phi$, $\rho\, d\theta$. Hence,

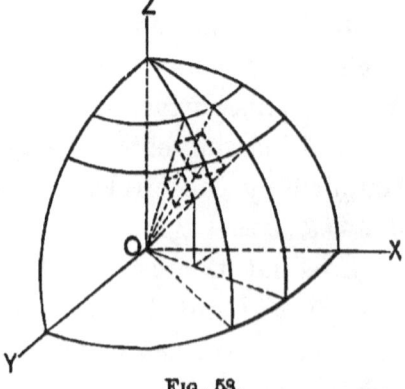

Fig. 58.

$$dM = \rho^2 \cos\theta\, d\rho\, d\phi\, d\theta, \quad x = \rho \cos\theta \cos\phi, \quad D = k\rho^2,$$

where k is the density at unit's distance from the center. Finally,

$$\bar{x} = \frac{\int_0^{\frac{\pi}{2}} \int_0^{\frac{\pi}{2}} \int_0^R k \cdot \rho^5 \cdot \cos^2\theta \cdot \cos\phi \cdot d\phi \cdot d\theta \cdot d\rho}{\int_0^{\frac{\pi}{2}} \int_0^{\frac{\pi}{2}} \int_0^R k \cdot \rho^4 \cdot \cos\theta \cdot d\phi \cdot d\theta \cdot d\rho} = \tfrac{5}{12}R.$$

PROBLEMS

1. Find the center of mass of a hemisphere.

2. Find the center of mass of the paraboloid of revolution generated by revolving $y^2 = 2px$ about the X-axis, and included by the planes $x = 0$, $x = x'$.

3. Find the center of mass of the solid generated by revolving half the ellipse $\dfrac{x^2}{a^2} + \dfrac{y^2}{b^2} = 1$ from $x = 0$ to $x = a$ about the X-axis.

4. Find the center of mass of the rectangular wedge.

ART. 62.—THEOREMS OF PAPPUS *

Multiplying both sides of the equation $\bar{y} = \dfrac{\int y \cdot ds}{s}$ by $2\pi s$, $2\pi \bar{y} \cdot s = 2\pi \int y \cdot ds$. The right-hand member of this equation represents the area of the surface generated by the revolution of the line s about the X-axis; the left-hand member is the length of the line s multiplied by the circumference of the circle described by the center of mass of the revolving line.

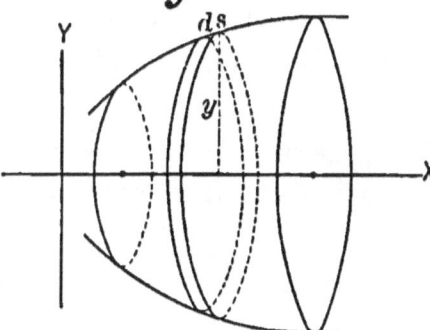

FIG. 59.

* First published by Pappus of Alexandria about the end of the third century of our era.

CENTER OF MASS AND MOMENT OF INERTIA

This furnishes a convenient determination of the center of mass of the line when the area generated is known, or of the area generated when the center of mass of the line is known.

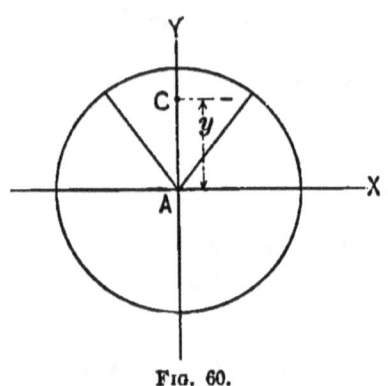

Fig. 60.

EXAMPLE I. — Determine the center of mass of the circular arc.

The arc revolving about the X-axis generates a zone whose area is $2\pi R \cdot \text{chord}$. Hence, $2\pi\bar{y} \cdot \text{arc} = 2\pi R \cdot \text{chord}$, and

$$\bar{y} = \frac{R \cdot \text{chord}}{\text{arc}},$$

as before found.

EXAMPLE II. — Find the area of the surface generated by revolving the cycloid $y = r\,\text{vers}^{-1}\frac{x}{r} + \sqrt{2rx - x^2}$ about its base.

The distance of the center of mass of the cycloidal arc from the base is $\frac{4}{3}r$, the length of the cycloid is $8r$. Hence, area $= 2\pi \cdot \frac{4}{3}r \cdot 8r = \frac{64}{3}\pi r^2$.

PROBLEMS

1. Find the area of the surface generated by the revolution about the X-axis of the circle of radius r, distance of center from X-axis a, where $a > r$.

2. Find the area of the surface generated by the revolution of a semi-circumference of radius r about a tangent at its middle point.

Multiplying both sides of the equation,

$$\bar{y} = \frac{\int\int y\,dx\,dy}{\int\int dx\,dy} = \frac{\frac{1}{2}\int y^2\,dx}{A} \text{ by } 2\pi A,\ 2\pi\bar{y} \cdot A = \pi\int y^2\,dx.$$

164   DIFFERENTIAL AND INTEGRAL CALCULUS

The right-hand member of this equation represents the volume of the solid generated by the revolution of the area A about the X-axis; the left-hand member is the area A multiplied by the circumference described by the center of mass of A. This furnishes a convenient determination of the center of mass of the area when the volume generated is known, or of the volume when the center of mass of the area is known.

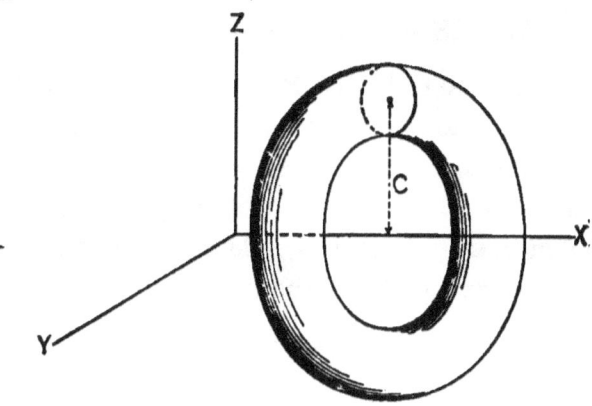

FIG. 61.

EXAMPLE I. — Find the volume generated by the revolution about the X-axis of the ellipse whose axes are $2a$ and $2b$, distance of center from X-axis c. $V = 2\pi c \cdot \pi ab = 2\pi^2 abc$.

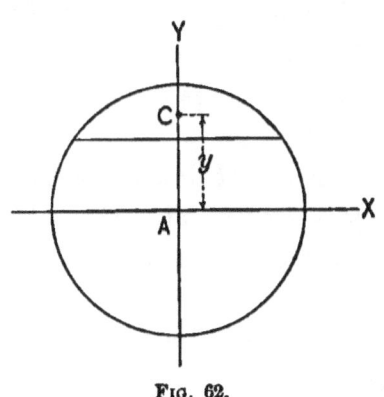

FIG. 62.

EXAMPLE II. — Find the center of mass of a circular sector.

The volume generated by the circular sector whose angle is $2\theta_0$, radius R, revolving about a diameter parallel to the chord of the sector is $2\pi R \cdot 2R\sin\theta_0 \cdot \tfrac{1}{3}R$, the area of the sector is $R^2 \cdot \theta_0$.

Hence, $\bar{y} = \dfrac{2}{3}\dfrac{R\sin\theta_0}{\theta_0}$.

CENTER OF MASS AND MOMENT OF INERTIA

PROBLEMS

1. Find the volume generated by revolving the cycloid $y = r \cdot \text{vers}^{-1}\frac{x}{r} + \sqrt{2rx - x^2}$ about its base.

2. Find the distance of the center of mass of half the ellipse $\frac{x^2}{a^2} + \frac{y^2}{b^2} = 1$ bounded by $x = 0$, $x = a$, from the center of the ellipse.

ART. 63. — MOMENT OF INERTIA

If the mass of a body is divided into infinitesimal elements of mass, and each element is multiplied by the square of its distance from a fixed line, the sum of all these products is called the moment of inertia of the body with respect to the straight line as axis. Denoting the infinitesimal element of mass by dm, its distance from the axis by y, and the moment of inertia by I, $I = \int y^2 \cdot dm$, the integration extending over the entire mass of the body.

Representing an infinitesimal element of the magnitude of the body by dM, the density of this element by D, $dm = D \cdot dM$, and the moment of inertia becomes $I = \int y^2 \cdot D \cdot dM$, the integration extending over the entire magnitude of the body.

The moment of inertia occurs very frequently in Strength of Materials and in the Theory of Rotation of Bodies. For example, in Problems 31, 32, 33 of Article 23, I is the moment of inertia of the cross-section of the beam.

If the material of the body is of uniform unit density, which is always assumed unless the contrary is stated, $D \cdot dM = dM$, and the moment of inertia depends only on the shape of the body.

The moment of inertia is one of the elements used in selecting shapes for engineering structures.

The moment of inertia divided by the mass of the body is called the square of the radius of gyration of the body for the moment axis used and is denoted by k^2. Hence $k^2 = \dfrac{\int y^2 \, dm}{m}$. k is the distance from the moment axis to the point where the mass of the body must be concentrated so that the moment of inertia of the concentrated mass shall equal the moment of inertia of the mass distributed throughout the body. In Problem 35 of Article 23 and in Problem 13 of Article 43, k^2 is the square of the radius of gyration of the turning body.

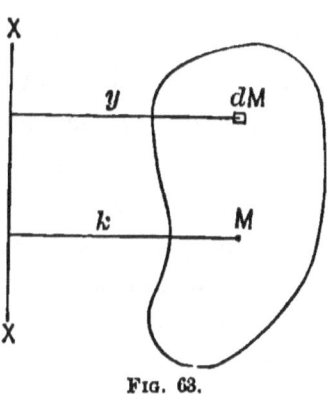

FIG. 63.

Denote by I_a the moment of inertia of a body for any moment axis AA, by I_c the moment of inertia of the same body for a parallel moment axis through the center of mass of the body. Through C, the center of mass of the body, pass a plane RS perpendicular to the line AC. Denote by D the distance between the two axes, by r the distance of dM from the axis AA, by r' the distance of dM from the axis CC. By definition $I_a = \int r^2 \cdot dM$, $I_c = \int r'^2 \cdot dM$. From the triangle ACP, $r^2 = r'^2 + D^2 + 2 r' D \sin \theta$. Hence

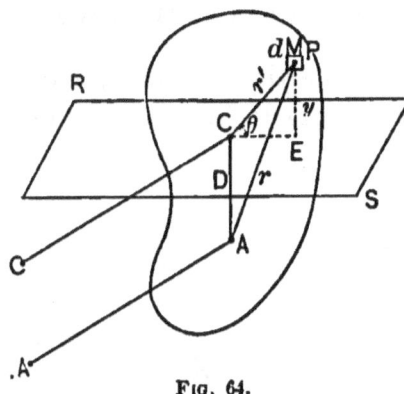

FIG. 64.

(1) $\quad \int r^2 \, dM = \int r'^2 \, dM + D^2 \int dM + 2 D \int r' \sin \theta \cdot dM.$

Now $r' \sin \theta = PE = y$ and $\int r' \sin \theta \cdot dM = \int y \, dM$.

But $\bar{y} = \dfrac{\int y \, dM}{M} = 0$, since $\bar{y}$ is the distance of the center of mass of the body from the plane RS, and this plane is drawn through C perpendicular to the line CA. Hence (1) becomes $I_a = I_c + D^2 \cdot M$, which may be written $I_c = I_a - D^2 \cdot M$, and is known as the reduction formula. Stated in words, this formula reads the moment of inertia of a body for any moment axis equals the moment of inertia for a parallel moment axis through the center of mass of the body plus the square of the distance between the two axes into the mass of the body.

Denoting the radii of gyration for the axis AA and the axis CC by K_a and K_c respectively, $I_a = M \cdot K_a^2$, $I_c = M \cdot K_c^2$, and by the reduction formula $K_a^2 = K_c^2 + D^2$.

ART. 64. — MOMENT OF INERTIA OF LINES AND SURFACES

EXAMPLE I. — Find the moment of inertia of a straight line of length l for moment axis perpendicular to line through one end of line.

Denoting an element of the line by dy, the distance of this element from the moment axis by y,

$$I = \int_0^l y^2 \cdot dy = \tfrac{1}{3} l^3.$$

The radius of gyration is found from

$$k^2 = \dfrac{\tfrac{1}{3} l^3}{l} = \tfrac{1}{3} l^2.$$

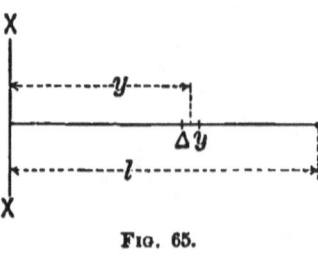

Fig. 65.

The moment of inertia for a parallel axis through the center of mass of the line is $I_c = \tfrac{1}{3} l^3 - \tfrac{1}{4} l^3 = \tfrac{1}{12} l^3$. The radius of gyration for this axis is found from $k_c^2 = \tfrac{1}{12} l^2$.

EXAMPLE II. — Find the moment of inertia of a triangle for moment axis through vertex parallel to base of triangle.

Breaking up the triangle into strips by lines parallel to the moment axis at intervals dy, calling the length of any strip x,

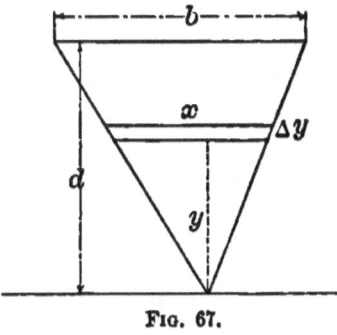

Fig. 67.

and its distance from the moment axis y, $dM = x\,dy$, and from similar triangles $\dfrac{x}{y} = \dfrac{b}{d}$. Hence,

$$I = \frac{b}{d}\int y^2\,dM = \frac{b}{d}\int_0^d y^3\,dy = \tfrac{1}{4}bd^3.$$

The radius of gyration is found from $\quad k^2 = \dfrac{\tfrac{1}{4}bd^3}{\tfrac{1}{2}bd} = \tfrac{1}{2}d^2.$

The moment of inertia for a parallel axis through the center of mass of triangle is

$$I_c = \tfrac{1}{4}bd^3 - \tfrac{1}{2}bd\,(\tfrac{2}{3}d)^2 = \tfrac{1}{36}bd^3 \text{ and } k_c^2 = \tfrac{1}{18}d^2.$$

The moment of inertia for the base of triangle as axis is

$$I_a = \tfrac{1}{36}bd^3 + \tfrac{1}{2}bd\,(\tfrac{1}{3}d)^2 = \tfrac{1}{12}bd^3 \text{ and } k_a^2 = \tfrac{1}{6}d^2.$$

EXAMPLE III. — Find the moment of inertia of a circle for axis through center perpendicular to plane of circle.

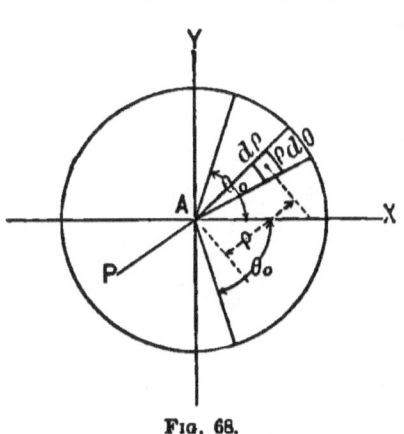

Fig. 68.

Here it is advisable to use polar coordinates. $dM = \rho\,d\theta\,d\rho$, and the distance of dM from the axis AP is ρ. Hence,

$$I_p = \int_0^{2\pi}\!\!\int_0^R \rho^3\,d\theta\,d\rho = \tfrac{1}{2}\pi R^4,$$

and $k_p^2 = \tfrac{1}{2}R^2$. I_p is called the polar moment of inertia of the circle.

Denoting by I_x the moment of

inertia of the circle for axis AX, by I_y the moment of inertia for axis AY,

$$I_x = \int y^2 dM, \quad I_y = \int x^2 dM,$$

and
$$I_x + I_y = \int (x^2 + y^2) dM = \int \rho^2 dM = I_p.$$

Since the circle is placed in exactly the same manner with respect to each of the diameters AX and AY, $I_x = I_y$. Hence,

$$2 I_x = I_p = \tfrac{1}{2} \pi R^4, \text{ and } I_x = \tfrac{1}{4} \pi R^4, \quad k_x^2 = \tfrac{1}{4} R^2.$$

PROBLEMS

1. Find moment of inertia of a rectangle base b, altitude d for base of rectangle as axis.

2. From Problem 1 find by reduction formula the moment of inertia of rectangle for axis through center of rectangle parallel to base.

3. Find moment of inertia of rectangle for axis through center perpendicular to plane of rectangle.

4. Find moment of inertia of isosceles triangle for axis of symmetry as moment axis.

5. Find moment of inertia of ellipse for major diameter as moment axis.

6. Find moment of inertia of ellipse for minor diameter as moment axis.

7. Find moment of inertia of ellipse for axis through center of ellipse perpendicular to plane of ellipse.

8. Find moment of inertia of parabolic segment for axis of parabola as moment axis.

9. Find moment of inertia of circular spandril for diameter as moment axis.

ART 65. — MOMENT OF INERTIA OF SOLIDS

EXAMPLE I. — Find the moment of inertia of the rectangular parallelopiped whose dimensions are b, d, h for axis through centers of two opposite faces.

Break up the solid into laminæ by planes perpendicular to the axis at intervals dx and call the distance of any lamina

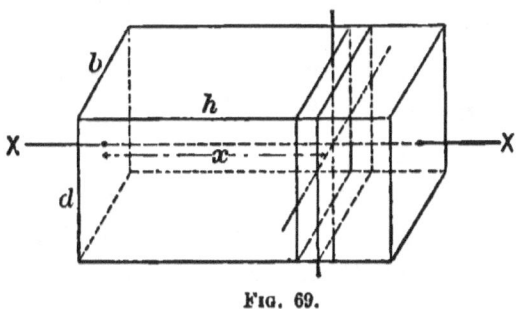

FIG. 69.

from one of the faces x. The lamina may be broken up into elements $dx\,dy\,dz$, and the moment of inertia of the lamina $= dx \int \int \rho^2 \, dy \, dz =$ moment of inertia of base of lamina multiplied by the thickness of the lamina. Hence the moment of inertia of the lamina is $\frac{1}{12}(b^2+d^2)bd\,dx$, and the moment of inertia of the parallelopiped is

$$I = \tfrac{1}{12}(b^2+d^2)bd \int_0^h dx = \tfrac{1}{12}(b^2+d^2)bdh.$$

The radius of gyration is found from

$$k^2 = \frac{\tfrac{1}{12}(b^2+d^2)bdh}{bdh} = \tfrac{1}{12}(b^2+d^2).$$

EXAMPLE II. — Find the moment of inertia of a cone of revolution for axis through center of mass of cone parallel to base of cone.

Break up the cone into laminæ by planes parallel to the base at intervals dy. Call the distance of any lamina from the vertex y, the radius of the base of the lamina x. The moment of inertia of the lamina for axis $X'X'$ through the center of the lamina and parallel to the axis XX is $\frac{1}{4}\pi x^4 dy$, the distance from $X'X'$ to XX is $\frac{3}{4}H - y$.

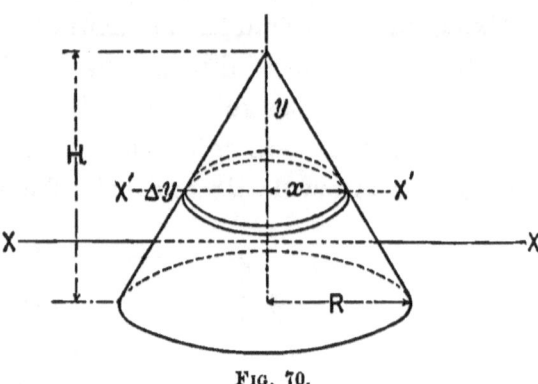

Fig. 70.

Hence by the reduction formula the moment of inertia of the lamina for axis XX is $\frac{1}{4}\pi x^4 dy + \pi x^2 (\frac{3}{4}H - y)^2 dy$, and the moment of inertia of the entire cone is

$$I = \int_0^H \{\tfrac{1}{4}\pi x^4 dy + \pi x^2 (\tfrac{3}{4}H - y)^2 dy\}.$$

By similar triangles $\dfrac{x}{y} = \dfrac{R}{H}$, hence

$$I = \int_0^H \left\{\frac{\pi}{4}\frac{R^4}{H^4}y^4 dy + \pi \frac{R^2}{H^2}(\tfrac{3}{4}H - y)^2 dy\right\} = \tfrac{1}{20}\pi R^2 H(R^2 + \tfrac{1}{4}H^2).$$

PROBLEMS

1. Find moment of inertia of the sphere for diameter as axis.

2. Find moment of inertia of cone of revolution for axis of symmetry as moment axis.

3. Find moment of inertia of cylinder of revolution for axis of symmetry as moment axis.

4. Find moment of inertia of cylinder of revolution for axis through center parallel to base.

5. Find moment of inertia of spherical cap for axis of symmetry as moment axis.

6. Find moment of inertia of ellipsoid for longest diameter as moment axis.

7. Find moment of inertia of the segment of a paraboloid of revolution for axis of symmetry as moment axis.

CHAPTER X

EXPANSIONS

Art. 66. — Convergent Power Series

The identical equation $(1-x)^2 \equiv 1 - 2x + x^2$ is true for all values of x.

Expanding the fraction $\dfrac{1}{1-x}$ into a series by division, there results the identical equation

$$\frac{1}{1-x} \equiv 1 + x + x^2 + x^3 + x^4 + \cdots x^{n-1} + x^n + x^{n+1} + x^{n+2} + x^{n+3} + \cdots,$$

where the number of terms in the series is infinite. This identity is not true for all values of x. For example, if $x = 2$, the fraction $\dfrac{1}{1-x}$ equals -1, the series equals infinity. To determine for what values of x the identity is true, denote the sum of the first n terms of the series by s_n, the sum of the remaining terms by r_n. Then $\dfrac{1}{1-x} \equiv s_n + r_n$, and if r_n can be made less than any assigned quantity, however small, by taking n sufficiently large, $\dfrac{1}{1-x} = \text{limit } s_n$ when limit $n = \infty$ and the series is said to be convergent. Now

$$r_n = x^n(1 + x + x^2 + x^3 + x^4 + \cdots).$$

When $x < 1$, $r_n = \dfrac{x^n}{1-x}$, and limit $r_n = 0$ when limit $n = \infty$. Hence when $x < 1$, the value of the fraction $\dfrac{1}{1-x}$ is correctly represented by the infinite series

(1) $\quad 1 + x + x^2 + x^3 + x^4 + \cdots x^{n-1} + x^n + x^{n+1} + x^{n+2} + \cdots,$

and s_n approaches the value of the fraction $\dfrac{1}{1-x}$ more and more closely the larger n is taken. For example, when $x = \tfrac{1}{2}$, the fraction equals 2, while $s_4 = \tfrac{15}{8} = 1.8889$, $s_8 = \tfrac{255}{128} = 1.9914$. This infinite series is convergent when $-1 < x < 1$ and converges towards 2 when $x = \tfrac{1}{2}$. The totality of values of x for which the infinite series is convergent is called the region of convergence of the infinite series.

FIG. 71.

The heavy portion of the straight line shows the region of convergence of the infinite series (1).

In general, the infinite series

$$a_0 + a_1 \cdot x + a_2 \cdot x^2 + a_3 \cdot x^3 + a_4 \cdot x^4 + \cdots$$
$$+ a_{n-1} \cdot x^{n-1} + a_n \cdot x^n + a_{n+1} \cdot x^{n+1} + \cdots,$$

where the coefficients are finite and independent of x, is called a power series. Denoting the sum of the first n terms of the power series by s_n, the sum of the remaining terms by r_n,

$$a_0 + a \cdot x + a_2 \cdot x^2 + a_3 \cdot x^3 + \cdots$$
$$+ a_{n-1} \cdot x^{n-1} + a_n \cdot x^n + a_{n+1} \cdot x^{n+1} + \cdots \equiv s_n + r_n.$$

For each value of x for which limit $r_n = 0$ when limit $n = \infty$, the power series is convergent and has a determinate finite value, which is the limit of s_n when limit $n = \infty$. Hence within its region of convergence the power series defines a function of x and may be denoted by $f(x)$.

A convenient test for the convergence of a power series is Cauchy's test, which reads:

A power series is convergent if from and after some fixed

term the ratio of each term to the preceding term is always numerically less than some number numerically less than unity.

For suppose that from and after the $m+2$ term, m being finite, of the power series

$$a_0 + a_1 \cdot x + a_2 \cdot x^2 + \cdots a_m \cdot x^m + a_{m+1} \cdot x^{m+1} + a_{m+2} \cdot x^{m+2} + \cdots$$

the ratio of each term to the preceding term is numerically less than r, and that $-1 < r < 1$. Denote the sum of the first m terms of the series by s_m, the sum of the next n terms by s_n, the sum of the remaining terms by r_n, the sum of the entire series by s. Then $s = s_m + s_n + r_n$,

$$s_n + r_n = a_m \cdot x^m + a_{m+1} \cdot x^{m+1} + a_{m+2} \cdot x^{m+2} + a_{m+3} \cdot x^{m+3} + \cdots$$
$$< a_m \cdot x^m (1 + r + r^2 + r^3 + \cdots r^n + r^{n+1} + r^{n+3} + \cdots)$$

by hypothesis. Hence $r_n < a_m \cdot x^m \dfrac{r^n}{1-r}$, and limit $r_n = 0$ when limit $n = \infty$, since a_m and x are by hypothesis not infinite and r is assumed less than unity. Hence the sum of the series $s = s_m +$ limit s_n when limit $n = \infty$; the series is convergent and Cauchy's test is proved.

For example, in the series $1 + x + x^2 + x^3 + \cdots x^n + x^{n+1} + \cdots$, the ratio of each term to the preceding term is x, and by Cauchy's test the series is convergent when $-1 < x < 1$.*

Art. 67. — Taylor's and Maclaurin's Series

The values of explicit algebraic functions can be directly calculated for arbitrarily assigned values of the independent variable. For example, if $y = x^3 - 7x + 7$, any value may be assigned to x, and the corresponding value of y becomes known.

* Euler seems to have been the first to call attention to the fact that infinite series can be safely used only when convergent.

Consider the function $f(x) = x^m$, when m is assumed to be a positive integer. When $x = b = a + h$, whence $h = b - a$,

$$f(b) \equiv (a+h)^m \equiv a^m + m \cdot a^{m-1} \cdot h + m(m-1)a^{m-2}\frac{h^2}{2!}$$
$$+ m(m-1)(m-2)a^{m-3}\frac{h^3}{3!} + \cdots$$

by the binomial formula. This result may be written,

(1) $f(b) \equiv f(a+h) \equiv f(a) + f'(a) \cdot h$
$$+ f''(a) \cdot \frac{h^2}{2!} + f'''(a) \cdot \frac{h^3}{3!} + \cdots$$

or (2) $f(b) \equiv f(a) + f'(a) \cdot (b-a)$
$$+ f''(a) \cdot \frac{(b-a)^2}{2!} + f'''(a) \cdot \frac{(b-a)^3}{3!} + \cdots,$$

where $f(a)$, $f'(a)$, $f''(a)$, $f'''(a)$, $\cdots$, are the values of the successive derivatives of $f(x) = x^m$ when $x = a$.

The symbol $n!$ stands for $1 \cdot 2 \cdot 3 \cdot 4 \cdot 5 \cdots n$, and is read factorial n.

It is proposed to derive a general series of the same form as series (1) and (2) for the approximate calculation of any function $f(x)$ for arbitrarily assigned values of the independent variable. The problem may be thus formulated: Let $f(x)$ and its successive derivatives $f'(x)$, $f''(x)$, $f'''(x)$, $\cdots$, be finite and continuous from $x = a$ to $x = b$, and denote by $f(a)$, $f'(a)$, $f''(a)$, $f'''(a)$, $\cdots$, the values of these functions when $x = a$. The value of $f(b)$ is to be found in terms of $f(a)$, $f'(a)$, $f''(a)$, $f'''(a)$, $\cdots$, and the powers of $b - a$.

The investigation is based on the following proposition, known as Rolle's theorem:

If the function $\phi(x)$ and its first derivative $\phi'(x)$ are finite and continuous from $x = a$ to $x = b$ and $\phi(a) = 0$ and $\phi(b) = 0$, the first derivative $\phi'(x)$ must vanish for some value of x between a and b.

EXPANSIONS

If $\phi'(x)$ is identically zero, the truth of the proposition is evident. If $\phi'(x)$ is not identically zero, $\phi'(x)$ must change sign between $x = a$ and $x = b$, for otherwise $\phi(x)$ would either continually increase or continually decrease from $x = a$ to $x = b$, and in neither case could $\phi(a)$ and $\phi(b)$ both be zero. Hence $\phi'(x)$ must change sign between $x = a$ and $x = b$. Since $\phi'(x)$ is assumed to be finite and continuous from $x = a$ to $x = b$, $\phi'(x)$ can change sign only by passing through zero. Therefore $\phi'(x)$ must vanish for some value of x between $x = a$ and $x = b$. Denoting this value of x by $a + \theta(b-a)$, where θ must be less than unity, $\phi'\{a + \theta(b-a)\} = 0$.

Now assume $f(b) \equiv f(a) + k_1(b-a)$, where k_1 is a constant to be determined. Write $\phi_1(x) \equiv f(b) - f(x) - k_1(b-x)$ and form the first derivation $\phi_1'(x) \equiv -f'(x) + k_1$. Since by hypothesis $\phi_1(x)$ and $\phi_1'(x)$ are finite and continuous from $x = a$ to $x = b$ and $\phi_1(a) = 0$ and $\phi_1(b) = 0$, $\phi_1'(x)$ must vanish for some value of x between a and b. Denoting this value of x by

$$a + \theta_1(b-a), \quad k_1 = f'\{a + \theta_1(b-a)\}.$$

Hence $\quad f(b) \equiv f(a) + f'\{a + \theta_1(b-a)\}(b-a)$.

Next assume $\quad f(b) \equiv f(a) + f'(a) \cdot (b-a) + k_2 \cdot \dfrac{(b-a)^2}{2!}$.

Write $\phi_2(x) \equiv f(b) - f(x) - f'(x) \cdot (b-x) - k_2 \cdot \dfrac{(b-x)^2}{2!}$ and form the first derivative $\phi_2'(x) \equiv -(b-x)f''(x) + k_2(b-x)$. Since by hypothesis $\phi_2(x)$ and $\phi_2'(x)$ are finite and continuous from $x = a$ to $x = b$ and $\phi_2(a) = 0$ and $\phi_2(b) = 0$, $\phi_2'(x)$ must vanish for some value of x between a and b. Denoting this value of x by $a + \theta_2(b-a)$, $k_2 = f''\{a + \theta_2(b-a)\}$. Hence

$$f(b) \equiv f(a) + f'(a) \cdot (b-a) + f''\{a + \theta_2(b-a)\} \cdot \dfrac{(b-a)^2}{2!}.$$

N

Assume
$$f(b)=f(a)+f'(a)\cdot(b-a)+f''(a)\cdot\frac{(b-a)^2}{2!}+k_3\cdot\frac{(b-a)^3}{3!}.$$

Write
$$\phi_3(x)\equiv f(b)-f(x)-f'(x)\cdot(b-x)$$
$$-f''(x)\cdot\frac{(b-x)^2}{2!}-k_3\cdot\frac{(b-x)^3}{3!},$$

and form the first derivative
$$\phi_3'(x)\equiv -2f'''(x)\cdot\frac{(b-x)^2}{2!}+k_3\cdot\frac{(b-x)^2}{2!}.$$

Since $\phi_3(x)$ and $\phi_3'(x)$ are by hypothesis finite and continuous from $x=a$ to $x=b$ and $\phi_3(a)=0$ and $\phi_3(b)=0$, $\phi_3'(x)$ must vanish for some value of x between a and b. Denoting this value of x by $a+\theta_3(b-a)$, $k_3=f'''\{a+\theta_3(b-a)\}$. Hence

$$f(b)\equiv f(a)+f'(a)\cdot(b-a)+f''(a)\cdot\frac{(b-a)^2}{2!}$$
$$+f'''\{a+\theta_3(b-a)\}\cdot\frac{(b-a)^3}{3!}.$$

Repeating this operation n times, there results

(1) $f(b)\equiv f(a)+f'(a)\cdot(b-a)+f''(a)\cdot\frac{(b-a)^2}{2!}+f'''(a)\cdot\frac{(b-a)^3}{3!}$
$$+f^{IV}(a)\cdot\frac{(b-a)^4}{4!}+\cdots f^{n-1}(a)\cdot\frac{(b-a)^{n-1}}{(n-1)!}$$
$$+f^n\{a+\theta(b-a)\}\cdot\frac{(b-a)^n}{n!},$$

where $0<\theta<1$.

The error committed by placing $f(b)$ equal to the sum of the first n terms of series (1) is $r_n=f^n\{a+\theta(b-a)\}\frac{(b-a)^n}{n!}$. If this error can be made indefinitely small by taking n sufficiently large, the series is convergent and can be used to calculate the value of $f(b)$ to any required degree of accuracy, provided $f(a)$, $f'(a)$, $f''(a)$, $f'''(a)$, $\cdots$ are known. When n

becomes indefinitely large, the series becomes an infinite series and the region of convergence is most readily determined by Cauchy's test.

In (1) place $b - a = h$, whence $b = a + h$. There results

(2) $f(b) \equiv f(a + h) \equiv f(a) + f'(a) \cdot h + f''(a) \cdot \dfrac{h^2}{2!} + f'''(a) \cdot \dfrac{h^3}{3!}$
$+ \cdots f^{n-1}(a) \cdot \dfrac{h^{n-1}}{(n-1)!} + f^n(a + \theta h) \cdot \dfrac{h^n}{n!}.$

This is Taylor's series. In (2) place $a = 0$. There results

(3) $f(h) \equiv f(0) + f'(0) \cdot h + f''(0) \cdot \dfrac{h^2}{2!} + f'''(0) \cdot \dfrac{h^3}{3!}$
$+ \cdots f^{n-1}(0) \cdot \dfrac{h^{n-1}}{(n-1)!} + f^n(\theta h) \cdot \dfrac{h^n}{n!}.$

This is Maclaurin's series. The only restrictions on a and h in these series are that $f(x)$, $f'(x)$, $f''(x)$, $f'''(x)$, $\cdots f^{n-1}(x)$ must be finite and continuous from $x = a$ to $x = a + h$ and that $f^n(a + \theta h) \cdot \dfrac{h^n}{n!}$ must become less than any assignable quantity when n is indefinitely increased. Since the quantities represented by a and h are not fixed, they may be denoted by x and y respectively, when Taylor's and Maclaurin's series become

$f(x + y) \equiv f(x) + f'(x) \cdot y + f''(x) \cdot \dfrac{y^2}{2!} + f'''(x) \cdot \dfrac{y^3}{3!} + \cdots$
$+ f^{n-1}(x) \cdot \dfrac{y^{n-1}}{(n-1)!} + f^n\{x + \theta(y - x)\} \cdot \dfrac{y^n}{n!};$

$f(y) \equiv f(0) + f'(0) \cdot y + f''(0) \cdot \dfrac{y^2}{2!} + f'''(0) \cdot \dfrac{y^3}{3!} + \cdots$
$+ f^{n-1}(0) \cdot \dfrac{y^{n-1}}{(n-1)!} + f^n(\theta y) \cdot \dfrac{y^n}{n!}.$ *

* Taylor (1685–1731) published his series in his "Methodus incrementorum." Maclaurin published his series in his "Treatise of Flexions" 1742. The expansions effected by these series had been previously obtained by laborious processes.

Taylor's series expands a function of the sum of two variables in the ascending powers of one of the variables; Maclaurin's formula expands a function of one variable in the ascending powers of that variable.

EXAMPLE I. — Expand $\log_e(1+y)$.

This expansion is effected by Maclaurin's series since $\log_e(1+y)$ is a function of one variable. Forming the successive derivatives,

$$f(y) = \log_e(1+y), \qquad f'''(y) = \frac{2!}{(1+y)^3},$$

$$f'(y) = \frac{1}{1+y}, \qquad f^{\text{IV}}(y) = \frac{-3!}{(1+y)^4}, \cdots,$$

$$f''(y) = \frac{-1}{(1+y)^2}, \qquad f^n(y) = \pm\frac{(n-1)!}{(1+y)^n},$$

hence $f(0) = 0$, $f'(0) = 1$, $f''(0) = -1$, $f'''(0) = 2!$,
$f^{\text{IV}}(0) = -3!, \cdots f^n(0) = \pm(n-1)!$.

Substituting in Maclaurin's series,

$$\log_e(1+y) \equiv y - \frac{y^2}{2} + \frac{y^3}{3} - \frac{y^4}{4} + \frac{y^5}{5} - \frac{y^6}{6} + \frac{y^7}{7} - \cdots$$

$$\pm \frac{y^{n-1}}{(n-1)!} \mp \frac{y^n}{n!}\frac{1}{(1+\theta y)^n}.$$

By Cauchy's test this series is convergent for values of y numerically less than unity. Hence this series may be used to calculate the Napierian logarithms of numbers from 0 to 2.

$$\begin{array}{ccc} -1 & 0 & +1 \end{array}$$

FIG. 72.

Taking $y = .5$ and $n = 13$, $\log_e 1.5 = .40546914$ with an error between .000000053 and .0000093.

The region of convergence of this expansion of $\log_e(1+y)$ is indicated graphically by the heavy line of the figure.

From this expansion of $\log_e(1+y)$ an expansion of $\log_e(1+z)$ with an enlarged region of convergence is obtained by the following analysis:

$$\log_e(1+y) \equiv y - \frac{y^2}{2} + \frac{y^3}{3} - \frac{y^4}{4} + \frac{y^5}{5} - \frac{y^6}{6} + \frac{y^7}{7} - \frac{y^8}{8} + \cdots$$

for $-1 < y < 1$.

$$\log_e(1-y) \equiv -y - \frac{y^2}{2} - \frac{y^3}{3} - \frac{y^4}{4} - \frac{y^5}{5} - \frac{y^6}{6} - \frac{y^7}{7} - \frac{y^8}{8} - \cdots$$

for $-1 < y < 1$.

By subtraction,

$$\log_e(1+y) - \log_e(1-y) \equiv \log_e \frac{1+y}{1-y} \equiv 2\left(y + \frac{y^3}{3} + \frac{y^5}{5} + \frac{y^7}{7} + \cdots\right)$$

for $-1 < y < 1$. Substituting

$$y = \frac{1}{2z+1}, \quad \frac{1+y}{1-y} = \frac{z+1}{z}, \text{ and when } -1 < y < 1, \ z > 0.$$

Hence $\log_e(1+z) = \log_e z$

$$+ 2\left(\frac{1}{2z+1} + \frac{1}{3(2z+1)^3} + \frac{1}{5(2z+1)^5} + \frac{1}{7(2z+1)^7} + \cdots\right)$$

for $z > 0$; a convenient formula for the calculation of the Napierian logarithms of numbers.

By Maclaurin's series

$$\log_a(1+y) \equiv \log_a e \left\{ y - \frac{y^2}{2} + \frac{y^3}{3} - \frac{y^4}{4} + \frac{y^5}{5} - \frac{y^6}{6} + \cdots \right\},$$

that is, $\log_a(1+y) \equiv \log_a e \cdot \log_e(1+y)$. Placing $1+y = a$,

$$\log_a a = \log_a e \cdot \log_e a,$$

whence $\log_a e = \dfrac{1}{\log_e a}$ and $\log_a(1+y) \equiv \dfrac{1}{\log_e a} \cdot \log_e(1+y)$.

The factor $\dfrac{1}{\log_e a}$, by which the Napierian logarithm of a number must be multiplied to obtain the logarithm of the same number in the system whose base is a, is called the modulus of the system of logarithms whose base is a. In the common system $a = 10$, and $\dfrac{1}{\log_e 10} = .43429448$.

Hence $\log_{10}(1+y) = .43429448 \log_e(1+y)$.

EXAMPLE II. — Expand $\log_e y$.

Here $f(y) = \log_e y$, $f'(y) = \dfrac{1}{y}$, $f''(y) = -\dfrac{1}{y^2}$, $f'''(y) = \dfrac{2}{y^3}$, $-\cdots$ $f^n(y) = \pm \dfrac{(n-1)!}{y^n}$. Hence, $f(0) = -\infty$, $f'(0) = \infty$, $f''(0) = -\infty$, $f'''(0) = \infty$, $\cdots$, $f^n(0) = \pm \infty$. The function $\log_e y$ cannot be expanded by Maclaurin's series into a power series in y. However, writing $\log y \equiv \log\{1 + (y-1)\}$, Example I. gives

$$\log y \equiv (y-1) - \frac{(y-1)^2}{2} + \frac{(y-1)^3}{3} - \frac{(y-1)^4}{4} + \frac{(y-1)^5}{5} - \cdots,$$

a power series in $y-1$ convergent when $0 < y < 2$.

EXAMPLE III. — Expand $(x+y)^m$, where m represents any finite number, positive or negative, integral or fractional.

This expansion is effected by Taylor's series.

Here $f(x) = x^m$, $f'(x) = mx^{m-1}$, $f''(x) = m(m-1)x^{m-2}$, $\cdots$

$$f^{n-1}(x) = m(m-1)\cdots(m-n+2)x^{m-n+1},$$

$$f^n(x) = m(m-1)\cdots(m-n+2)(m-n+1)x^{m-n}, \cdots$$

Substituting in Taylor's series,

$$(x+y)^m \equiv x^m + m \cdot x^{m-1} \cdot y + \frac{m(m-1)}{2!}x^{m-2}y^2$$

$$+ \frac{m(m-1)(m-2)}{3!}x^{m-3}y^3 + \cdots$$

$$\frac{m(m-1)\cdots(m-n+2)}{(n-1)!}x^{m-n+1}y^{n+1}$$

$$+ \frac{m(m-1)\cdots(m-n+2)(m-n+1)}{n!}x^{m-n}y^n + \cdots.$$

The ratio of the nth term to the $(n-1)$th term is $r = \left(\frac{m+1}{n} - 1\right)\frac{y}{x}$. Since m is finite, the factor $\frac{m+1}{n} - 1$ approaches unity when n is indefinitely increased. Hence if $\frac{y}{x} < 1$, the ratio r becomes less than unity when the series is sufficiently extended, and the series is convergent by Cauchy's test. This proves the binomial theorem for all finite exponents, provided $-1 < \frac{y}{x} < 1$.

EXAMPLE IV. — Expand $(2 - x + y)^{-\frac{1}{2}}$ in ascending powers of y.

Substituting $v = 2 - x$, $(2 - x + y)^{-\frac{1}{2}}$ becomes $(v + y)^{-\frac{1}{2}}$. By Taylor's series,

$$(v+y)^{-\frac{1}{2}} \equiv v^{-\frac{1}{2}} + \frac{d}{dv}(v^{-\frac{1}{2}}) \cdot y + \frac{d^2}{dv^2}(v^{-\frac{1}{2}}) \cdot \frac{y^2}{2!} + \frac{d^3}{dv^3}(v^{-\frac{1}{2}}) \cdot \frac{y^3}{3!} + \cdots$$

$$\equiv v^{-\frac{1}{2}} - \tfrac{1}{2} v^{-\frac{3}{2}} \cdot y + \tfrac{3}{8} v^{-\frac{5}{2}} \cdot y^2 - \tfrac{5}{16} v^{-\frac{7}{2}} \cdot y^3 + \cdots ;$$

whence, $(2 - x + y)^{-\frac{1}{2}} \equiv (2 - x)^{-\frac{1}{2}} - \tfrac{1}{2}(2 - x)^{-\frac{3}{2}} \cdot y$
$+ \tfrac{3}{8}(2 - x)^{-\frac{5}{2}} \cdot y^2 - \tfrac{5}{16}(2 - x)^{-\frac{7}{2}} \cdot y^3 + \cdots,$

convergent when $-1 < \frac{y}{2-x} < 1$.

PROBLEMS

Expand and determine the region of convergence,

1. $\sin y$.
2. $\cos y$.
3. $\sin(x + y)$.
4. $\cos(x + y)$.
5. $\sin(x - y)$.
6. $\cos(x - y)$.
7. e^y.
8. e^{-y}.
9. $\log(x + y)$.
10. $\tan^{-1} x$.
11. $\sin^{-1} x$.
12. $\log\{x + \sqrt{1 + x^2}\}$.
13. $\sinh x \equiv \tfrac{1}{2}(e^x - e^{-x})$.
14. $\cosh x \equiv \tfrac{1}{2}(e^x + e^{-x})$.
15. $e^x \sin x$.
16. $e^{-\sin^{-1} x}$.
17. $(1 - \sin^2 x)^{-\frac{1}{2}}$.
18. $\tan x$.
19. $\log \cos x$.
20. $\cot y$.
21. $(1 - x + y)^{\frac{1}{2}}$.
22. $(x^2 - y^3)^{-1}$.
23. $\log(x^3 - y^2)$.

24. Compute $\sin 10°$. Here $x = \dfrac{10°}{57°.3}$.

25. Compute $\sinh .2$.

26. Compute $\log_e 2$ and $\log_{10} 2$.   **27.** Expand e^{-z^2}.

Art. 68. — Euler's Formulas for Sine and Cosine

Let the series $e^z \equiv 1 + z + \dfrac{z^2}{2!} + \dfrac{z^3}{3!} + \dfrac{z^4}{4!} + \dfrac{z^5}{5!} + \dfrac{z^6}{6!} + \cdots$, which is the development by Maclaurin's formula of e^z when z is real, be adopted as the definition of e^z for all values of z, real and complex. Placing $z = ix$, where i stands for $\sqrt{-1}$, whence $i^{4n} = 1$, $i^{4n+1} = i$, $i^{4n+2} = -1$, $i^{4n+3} = -i$ for all integral values of n,

(1) $\quad e^{iz} \equiv 1 + ix - \dfrac{x^2}{2!} - i\dfrac{x^3}{3!} + \dfrac{x^4}{4!} + i\dfrac{x^5}{5!} - \dfrac{x^6}{6!} - i\dfrac{x^7}{7!} + \cdots$.

In like manner, placing $z = -ix$,

(2) $\quad e^{-iz} \equiv 1 - ix - \dfrac{x^2}{2!} + i\dfrac{x^3}{3!} + \dfrac{x^4}{4!} - i\dfrac{x^5}{5!} - \dfrac{x^6}{6!} + i\dfrac{x^7}{7!} + \cdots$.

Taking half the sum of (1) and (2),

(3) $\quad \dfrac{e^{iz} + e^{-iz}}{2} \equiv 1 - \dfrac{x^2}{2!} + \dfrac{x^4}{4!} - \dfrac{x^6}{6!} + \dfrac{x^8}{8!} - \cdots \equiv \cos x$;

dividing half the difference of (1) and (2) by $2i$,

(4) $\quad \dfrac{e^{iz} - e^{-iz}}{2i} \equiv x - \dfrac{x^3}{3!} + \dfrac{x^5}{5!} - \dfrac{x^7}{7!} + \dfrac{x^9}{9!} - \cdots \equiv \sin x$.

In general, $\sin(nx) = \dfrac{1}{2i}\{e^{inx} - e^{-inx}\}$, $\cos(nx) = \tfrac{1}{2}\{e^{inx} + e^{-inx}\}$.

These results are known as Euler's formulas for sine and cosine.

These formulas are useful in trigonometric transformations

and in integrating derivatives involving trigonometric and exponential functions.

Since the i in $u = ix$ is of the nature of a constant factor, $\dfrac{du}{dx} = i$.

EXAMPLE I. — Find the value of $\cos^5 x$ in terms of the functions of the multiples of x.

$$\cos^5 x = \{\tfrac{1}{2}(e^{ix} + e^{-ix})\}^5$$
$$= \tfrac{1}{32}\{e^{i5x} + 5\,e^{i3x} + 10\,e^{ix} + 10\,e^{-ix} + 5\,e^{-i3x} + e^{-i5x}\}$$
$$= \tfrac{1}{16}\{\tfrac{1}{2}(e^{i5x} + e^{-i5x}) + \tfrac{5}{2}(e^{i3x} + e^{-i3x}) + \tfrac{10}{2}(e^{ix} + e^{-ix})\}$$
$$= \tfrac{1}{16}\{\cos(5x) + 5\cos(3x) + 10\cos x\}.$$

EXAMPLE II. — Integrate $\dfrac{dy}{dx} = \sin^2 x \cdot \cos^2 x$.

Substituting for $\sin x = \dfrac{1}{2i}(e^{ix} - e^{-ix})$ for $\cos x = \tfrac{1}{2}(e^{ix} + e^{-ix})$,

$$\dfrac{dy}{dx} = -\tfrac{1}{16}(e^{i4x} + e^{-i4x} - 2) = -\tfrac{1}{8}\cos(4x) + \tfrac{1}{8}. \quad \text{Hence}$$
$$y = -\tfrac{1}{32}\sin(4x) + \tfrac{1}{8}x + C.$$

EXAMPLE III. — Integrate $\dfrac{dy}{dt} = e^{-Pt}\sin(pt)$, where P and p are constants.

Substituting $\sin(pt) = \dfrac{1}{2i}(e^{ipt} - e^{-ipt})$,

$$\dfrac{dy}{dt} = \dfrac{1}{2i}\{e^{(-P+ip)t} - e^{(-P-ip)t}\},$$

hence $\quad y = \dfrac{1}{2i}\left\{\dfrac{1}{-P+ip}e^{(-P+ip)t} + \dfrac{1}{P+ip}e^{(-P-ip)t}\right\} + C$

$$= -\dfrac{1}{2i}\dfrac{e^{-Pt}}{P^2+p^2}\{P(e^{ipt} - e^{-ipt}) + ip(e^{ipt} + e^{-ipt})\} + C$$

$$= -\dfrac{e^{-Pt}}{P^2+p^2}\{P\sin(pt) + p\cos(pt)\} + C.$$

PROBLEMS

Find in terms of the functions of the multiples of x,

1. $\sin^4 x$.
2. $\sin^7 x$.
3. $\sin^2 x \cdot \cos^2 (2x)$.

Integrate, 4. $\dfrac{dy}{dx} = \sin^3 x$.   5. $\dfrac{dy}{dx} = \sin^4 x$.

6. $\dfrac{dy}{dx} = \cos^2 x$.   7. $\dfrac{dy}{dx} = \sin^6 x$.   8. $\dfrac{dy}{dx} = e^{2x} \cdot \sin^2 x$.

9. Show that $e^{ix} = \cos x + i \sin x$, $e^{-ix} = \cos x - i \sin x$.

Art. 69. — Differentiation and Integration of Power Series

Consider the power series,

(1) $\quad f(x) = a_0 + a_1 \cdot x + a_2 \cdot x^2 + \cdots$
$$a_n \cdot x^n + a_{n+1} \cdot x^{n+1} + a_{n+2} \cdot x^{n+2} + \cdots,$$

and denote the sum of the first n terms by $s_n(x)$, the sum of the remaining terms by $r_n(x)$.

Absolute Convergence. — Denote the numerical or absolute value of any quantity z by the notation $|z|$. So that $|-5| = 5$, $|\tfrac{2}{3} - 1| = \tfrac{1}{3}$.

Suppose in the power series (1), denoting $|a_n|$ by A_n, that $A_n X_0^n < M$, where M is a finite quantity, for all values of n.

$$\underline{\qquad\qquad -X_0 \qquad\qquad O \qquad\qquad +X_0 \qquad\qquad}$$
Fig. 73.

Take $|x| = X < X_0$; then, since $A_n < \dfrac{M}{X_0^n}$ by hypothesis,

$$A_0 + A_1 X + A_2 X^2 + \cdots A_n X^n + \cdots$$
$$< M + M\frac{X}{X_0} + M\left(\frac{X}{X_0}\right)^2 + \cdots M\left(\frac{X}{X_0}\right)^n + \cdots.$$

Since by hypothesis M is finite and $\dfrac{X}{X_0} < 1$, by Cauchy's test the right-hand member of the inequality, and consequently the left-hand member, is a convergent series. That is, the sum of the absolute values of the terms of the original power series is convergent for $|x| < X_0$ if $A_n X_0^n < M$. This is expressed by saying that the given power series is absolutely convergent within the region $-X_0 < x < X_0$.

EXAMPLE. — In the expansion,
$$(1-x^2)^{-\tfrac{1}{2}} \equiv 1 + \frac{x^2}{2} + \tfrac{3}{8}x^4 + \tfrac{5}{16}x^6 + \tfrac{35}{128}x^8 + \tfrac{35}{144}x^{10} + \cdots,$$

when $x = 1$, each term is less than 2. Hence, the series is absolutely convergent for $-1 < x < 1$.

Uniform Convergence. — Writing

$$A_0 + A_1 X + A_2 X^2 + \cdots A_n X^n + A_{n+1} X^{n+1} + \cdots \equiv S_n(X) + R_n(X),$$

$$r_n(x) \not> R_n(x) < M\left(\frac{X}{X_0}\right)^n + M\left(\frac{X}{X_0}\right)^{n+1} + \cdots \text{ for } |x| = X < X_0.$$

But $M\left(\dfrac{X}{X_0}\right)^n + M\left(\dfrac{X}{X_0}\right)^{n+1} + M\left(\dfrac{X}{X_0}\right)^{n+2} + \cdots = M\left(\dfrac{X}{X_0}\right)^n \dfrac{1}{1-\dfrac{X}{X_0}}.$

Now, n may be taken so large that, for all values of $X < X_0$, $M\left(\dfrac{X}{X_0}\right)^n \dfrac{1}{1-\dfrac{X}{X_0}}$ for the same value of n becomes less than ϵ, however small ϵ may be assumed. Consequently, since $r_n(x) \not> R_n(X)$, for all values of x between $-X_0$ and $+X_0$ the expression $r_n(x)$ can be made less than ϵ for one and the same value of n. This fact is expressed by saying that the power series is uniformly convergent in the region $-X_0 < x < X_0$.

EXAMPLE.—The power series $1 + x + x^2 + x^3 + \cdots x^n + x^{n-1} + \cdots$ is absolutely convergent for $|x| < 1$. Assuming $\epsilon = .000001$, determine n so that for this value of n $r_n(x) < \epsilon$ for all values of x between $-\tfrac{1}{2}$ and $+\tfrac{1}{2}$.

Since $r_n = \dfrac{x^n}{1-x}$, by the conditions of the problem $r_n < (\tfrac{1}{2})^{n-1}$. The conditions of the problem are satisfied when $(\tfrac{1}{2})^{n-1} = .000001$, that is, when $n = 22$.

Continuity. — Denote by x and x_0 any two quantities numerically less than X_0. Since $f(x) = s_n(x) + r_n(x)$,

$$f(x) - f(x_0) = \{s_n(x) - s_n(x_0)\} + \{r_n(x) - r_n(x_0)\}.$$

Since n, however large, is supposed to be finite, x may be taken sufficiently near x_0 to make $|s_n(x) - s_n(x_0)| < \epsilon$, however small ϵ may be assumed; and since by hypothesis x and x_0 are within the region of uniform convergence of the power series, n may be taken so large that $|r_n(x)|$ and $|r_n(x_0)|$ each become less than ϵ. Consequently $|f(x) - f(x_0)| < 3\epsilon$, and the function defined by the power series is continuous in the region of uniform convergence.

Integration. — To show that between limits within the region of uniform convergence the limit of the sum of the integrals of the terms of a power series is the integral of the limit of the power series, write

$$\int_\alpha^\beta f(x)\,dx \equiv \int_\alpha^\beta a_0 \cdot dx + \int_\alpha^\beta a_1 \cdot x\,dx + \int_\alpha^\beta a_2 \cdot x^2\,dx + \cdots$$
$$+ \int_\alpha^\beta r_n(x) \cdot dx,$$

where α and β lie within the region of uniform convergence of $f(x) \equiv a_0 + a_1 \cdot x + a_2 \cdot x^2 + a_3 \cdot x^3 + \cdots$

Now $\int_\alpha^\beta r_n(x) \cdot dx < \epsilon \int_\alpha^\beta dx = \epsilon(\beta - \alpha)$, where ϵ is a quantity which approaches zero as n approaches infinity. This proves the proposition.

The series
$$f(x) \equiv a_0 \cdot \psi(x) + a_1 \cdot \psi(x) \cdot \phi(x) + a_2 \cdot \psi(x) \cdot [\phi(x)]^2 + \cdots$$
may be integrated term by term between the limits $x = \alpha$ and $x = \beta$ provided α and β lie within the region of uniform convergence of the series $a_0 + a_1 \cdot \phi(x) + a_2 \cdot [\phi(x)]^2 + \cdots$ and $\psi(x)$ is finite from $x = \alpha$ to $x = \beta$.

EXAMPLE I. — Expand $\tan^{-1} x$ into a power series by integration.
$$\frac{d}{dx} \tan^{-1} x = \frac{1}{1+x^2} \equiv 1 - x^2 + x^4 - x^6 + x^8 - x^{10} + \cdots,$$
a power series uniformly convergent for $|x| < 1$. Hence term by term integration gives a valid result, and
$$\tan^{-1} x \equiv x - \frac{x^3}{3} + \frac{x^5}{5} - \frac{x^7}{7} + \frac{x^9}{9} - \frac{x^{11}}{11} + \cdots \text{ for } |x| < 1.$$

From this expansion is obtained Euler's series for the calculation of π. Writing
$$\tan u = \tfrac{1}{2} \text{ and } \tan v = \tfrac{1}{3}, \ \tan(u+v) = 1 = \tan\frac{\pi}{4}.$$

Hence $\quad \dfrac{\pi}{4} = u + v = \tan^{-1}\tfrac{1}{2} + \tan^{-1}\tfrac{1}{3},$

and
$$\frac{\pi}{4} = \left(\frac{1}{2} - \frac{1}{3 \cdot 2^3} + \frac{1}{5 \cdot 2^5} - \cdots\right) + \left(\frac{1}{3} - \frac{1}{3 \cdot 3^3} + \frac{1}{5 \cdot 3^5} - \cdots\right)$$
$$= \left(\frac{1}{2} + \frac{1}{3}\right) - \frac{1}{3}\left(\frac{1}{2^3} + \frac{1}{3^3}\right) + \frac{1}{5}\left(\frac{1}{2^5} + \frac{1}{3^5}\right) - \cdots.$$

EXAMPLE II. — Find the value of t when
$$t = \frac{4r}{\sqrt{2g}} \int_0^h \frac{dy}{\sqrt{(h-y)(2ry - y^2)}}, \ h < 2r.$$
Here t is the time of vibration of a simple pendulum of length r, the bob starting at a distance h above the horizontal through its lowest position.

Substituting $y = h \cdot \sin^2 \theta$,

$$t = \frac{4r}{\sqrt{2g}} \int_0^{\frac{\pi}{2}} \frac{2h \cdot \sin\theta \cdot \cos\theta \cdot d\theta}{\sqrt{h} \cdot \cos\theta \cdot \sin\theta \sqrt{2rh - h^2 \sin^2\theta}}$$

$$= 4\sqrt{\frac{r}{g}} \int_0^{\frac{\pi}{2}} \frac{d\theta}{\sqrt{\left(1 - \frac{h}{2r}\sin^2\theta\right)}} = 4\sqrt{\frac{r}{g}} \int_0^{\frac{\pi}{2}} \left(1 - \frac{h}{2r}\sin^2\theta\right)^{-\frac{1}{2}} d\theta$$

$$= 4\sqrt{\frac{r}{g}} \int_0^{\frac{\pi}{2}} \left[1 + \frac{1}{2} \cdot \frac{h}{2r} \cdot \sin^2\theta + \frac{1}{2} \cdot \frac{3}{4} \cdot \left(\frac{h}{2r}\right)^2 \cdot \sin^4\theta \right.$$

$$\left. + \frac{1}{2} \cdot \frac{3}{4} \cdot \frac{5}{6} \cdot \left(\frac{h}{2r}\right)^3 \cdot \sin^6\theta + \cdots \right] d\theta$$

$$= 4\sqrt{\frac{r}{g}} \frac{\pi}{2} \left[1 + \left(\frac{1}{2}\right)^2 \cdot \frac{h}{2r} + \left(\frac{1}{2} \cdot \frac{3}{4}\right)^2 \cdot \left(\frac{h}{2r}\right)^2 \right.$$

$$\left. + \left(\frac{1}{2} \cdot \frac{3}{4} \cdot \frac{5}{6}\right)^2 \cdot \left(\frac{h}{2r}\right)^3 + \cdots \right].$$

If h is small compared with $2r$, $t = 2\pi\sqrt{\frac{r}{g}}$ is an approximation sufficiently accurate for most purposes.

Differentiation. — To find under what conditions the sum of the derivatives of the terms of a power series,

$$f(x) \equiv a_0 + a_1 \cdot x + a_2 \cdot x^2 + a_3 \cdot x^3 + \cdots a_n \cdot x^n + \cdots,$$

uniformly convergent for $|x| < X_0$, is the derivative of the function defined by the power series, write

$$\phi(x) \equiv a_1 + 2a_2 \cdot x + \cdots n \cdot a_n \cdot x^{n-1} + (n+1) \cdot a_{n+1} \cdot x^n + \cdots.$$

By hypothesis the series defining $f(x)$ is uniformly convergent for $|x| < X_0$. If from and after some fixed term the ratio of the corresponding terms of $\phi(x)$ and $f(x)$ is not greater than unity, $\phi(x)$ is uniformly convergent when $f(x)$ is uniformly convergent. The ratio of the $(n+1)$th terms of $\phi(x)$ and $f(x)$ is $n\left(\dfrac{x_\phi}{x_f}\right)^n \dfrac{1}{x_\phi}$, where x_ϕ and x_f denote respectively the variables of the ϕ and f series. Hence $\phi(x)$ is uniformly convergent when $\left|n\left(\dfrac{x_\phi}{x_f}\right)^n \dfrac{1}{x_\phi}\right| \not> 1$, that is, when $n\left|\dfrac{x_\phi}{x_f}\right|^n \not> |x_\phi|$. If x_ϕ is taken less than x_f, limit $n\left(\dfrac{x_\phi}{x_f}\right)^n = 0$ when limit $n = \infty$. Hence $\phi(x)$ is uniformly convergent for $|x| < X$, when X lies within the region of uniform convergence of $f(x)$. For these values of x, $\phi(x)$ may be integrated term by term, and $f(x) - \int_\alpha^\beta \phi(x)\,dx \equiv a_0 + a_1 \cdot \alpha + a_2 \cdot \alpha^2 + a_3 \cdot \alpha^3 + \cdots$, when α and β lie within the region of uniform convergence. Differentiating this result, $f'(x) - \phi(x) \equiv 0$; hence

$$f'(x) \equiv a_1 + 2a_2 \cdot x + \cdots + n \cdot a_n \cdot x^{n-1} + \cdots.$$

That is, a uniformly convergent power series may be differentiated term by term as long as x is within the region of uniform convergence of the power series.

EXAMPLE. — The expansion

(1) $(1-x^2)^{-\frac{1}{2}} \equiv 1 + \tfrac{1}{2} x^2 + \tfrac{1}{2}\cdot\tfrac{3}{4} x^4 + \tfrac{1}{2}\cdot\tfrac{3}{4}\cdot\tfrac{5}{6} x^6 + \tfrac{1}{2}\cdot\tfrac{3}{4}\cdot\tfrac{5}{6}\cdot\tfrac{7}{8} x^8 + \cdots$

is uniformly convergent for $|x| < 1$. Obtain the expansion of $(1-x^2)^{-\frac{3}{2}}$ and $(1-x^2)^{-\frac{5}{2}}$ by differentiation.

Differentiating (1) and dividing by x,

(2) $(1-x^2)^{-\frac{3}{2}} \equiv 1 + \tfrac{3}{2} x^2 + \tfrac{15}{8} x^4 + \tfrac{35}{16} x^6 + \cdots;$

differentiating (2) and dividing by $3x$,

(3) $(1-x^2)^{-\frac{5}{2}} \equiv 1 + \tfrac{5}{2} x^2 + \tfrac{35}{8} x^4 + \cdots.$

PROBLEMS

Expand by integration,

1. $\log(1+x)$. 2. $\log(1-x)$. 3. $\sin^{-1} x$.

From the expansion of $(1-x)^{-1}$ obtain by differentiation the expansion of,

4. $(1-x)^{-2}$. 5. $(1-x)^{-3}$. 6. $(1-x)^{-4}$.

7. Find the length of the ellipse $x = a \cos \phi$, $y = b \sin \phi$.

If the arc is measured from the end of the major axis,

$$s = a \int_0^\phi \sqrt{1 - e^2 \sin^2 \phi}\, d\phi,$$

where e is the eccentricity of the ellipse.

The entire length is $4 a \int_0^{\frac{\pi}{2}} \sqrt{1 - e^2 \sin^2 \phi}\, d\phi$.

8. The discharge of water per second through a circular orifice of radius r, the plane of the orifice being vertical, when h is the head of water on the center of the orifice, is $Q = 2 \int_{-r}^{+r} \sqrt{r^2 - y^2} \sqrt{2g(h-y)}\, dy$. Find Q.

9. Evaluate the definite integral $\int_{-h}^{+h} e^{-x^2} \cdot dx$. This is called the probability integral.

10. Evaluate $\int_0^1 \frac{e^x - e^{-x}}{x}\, dx$. Expand e^x and e^{-x} separately, and express $\frac{e^x - e^{-x}}{x}$ in the form of an infinite series.

ART. 70. — EXPANSION OF $u_1 \equiv f(x + h, y + k)$

Let $u \equiv f(x, y)$ denote a continuous function of two independent variables. Denote by $u_1 \equiv f(x + h, y + k)$ the value of u when x and y are increased by h and k respectively. u_1 is to be expanded into an equivalent series in the ascending powers of h and k.

Denoting by u_0 the value of u when x is increased by h and y remains unchanged, by Taylor's series,

$$u_0 \equiv f(x + h, y) \equiv u + \frac{\partial u}{\partial x} \cdot h + \frac{\partial^2 u}{\partial x^2} \cdot \frac{h^2}{2!} + \frac{\partial^3 u}{\partial x^3} \cdot \frac{h^3}{3!} + \frac{\partial^4 u}{\partial x^4} \cdot \frac{h^4}{4!} + \cdots.$$

EXPANSIONS

Now u_1 is the value of u_0 when y is increased by k, x remaining unchanged. Hence $u_1 \equiv f(x+h, y+k)$

$$\equiv u_0 + \frac{\partial u_0}{\partial y} \cdot k + \frac{\partial^2 u_0}{\partial y^2} \cdot \frac{k^2}{2!} + \frac{\partial^3 u_0}{\partial y^3} \cdot \frac{k^3}{3!} + \frac{\partial^4 u_0}{\partial y^4} \cdot \frac{k^4}{4!} + \cdots$$

$$\equiv u + \frac{\partial u}{\partial x} \cdot h + \frac{\partial^2 u}{\partial x^2} \cdot \frac{h^2}{2!} + \frac{\partial^3 u}{\partial x^3} \cdot \frac{h^3}{3!} + \frac{\partial^4 u}{\partial x^4} \cdot \frac{h^4}{4!} + \cdots$$

$$+ \frac{\partial u}{\partial y} \cdot k + \frac{\partial^2 u}{\partial x \partial y} \cdot hk + \frac{\partial^3 u}{\partial x^2 \partial y} \cdot \frac{h^2 k}{2!} + \frac{\partial^4 u}{\partial x^3 \partial y} \cdot \frac{h^3 k}{3!} + \cdots$$

$$+ \frac{\partial^2 u}{\partial y^2} \cdot \frac{k^2}{2!} + \frac{\partial^3 u}{\partial x \partial y^2} \cdot \frac{hk^2}{2!} + \frac{\partial^4 u}{\partial x^2 \partial y^2} \cdot \frac{h^2 k^2}{2!} + \cdots$$

$$+ \frac{\partial^3 u}{\partial y^3} \cdot \frac{k^3}{3!} + \frac{\partial^4 u}{\partial x \partial y^3} \cdot \frac{hk^3}{3!} + \cdots$$

$$+ \frac{\partial^4 u}{\partial y^4} \cdot \frac{k^4}{4!} + \cdots.$$

For example, in the sphere $x^2 + y^2 + z^2 = 25$, at the point $(0+h,\ 4+k,\ z)$, $z = 3 - \frac{4}{3}k - \frac{1}{3}h^2 - \frac{1}{3}k^2 - \cdots$. If $h = .5$ and $k = .1$, $z = 2.78$.

o

CHAPTER XI

APPLICATIONS OF TAYLOR'S SERIES

Art. 71. — Maxima and Minima by Expansion

If the function $y = f(x)$ is continuous in the neighborhood of (x_0, y_0), and y_1 denotes the value of y corresponding to $x = x_0 \pm h$, by Taylor's series,

$$(1) \quad y_1 - y_0 = f'(x_0)(\pm h) + f''(x_0)\frac{(\pm h)^2}{2!} + f'''(x_0)\frac{(\pm h)^3}{3!} + \cdots.$$

If h approaches zero, $y_1 - y_0$ approaches the term of the right-hand number of (1) which contains the lowest power of h. Hence, if $f'(x_0) \neq 0$, $y_1 - y_0$ changes sign with h, and y_0 is neither a maximum nor a minimum; if $f'(x) = 0$ and $f''(x_0)$ is negative, $y - y_0$ is negative for $+h$ and $-h$, and y_0 is a maximum; if $f'(x_0) = 0$ and $f''(x_0)$ is positive, $y_1 - y_0$ is positive for $+h$ and $-h$, and y_0 is a maximum; if $f'(x_0) = 0$, $f''(x_0) = 0$ and $f'''(x_0) \neq 0$, $y_1 - y_0$ changes sign with h, and y_0 is neither a maximum nor a minimum.

In general, if the first derivative in the expansion (1) not to vanish is of an odd order, the function is neither at a maximum nor at a minimum; if the first derivative not to vanish is of an even order, the function is at a maximum if this derivative is negative, at a minimum if this derivative is positive. This agrees with the results of Art. 24.

APPLICATIONS OF TAYLOR'S SERIES

If the function $u = F(x, y)$ is continuous in the neighborhood of (x_0, y_0), and u_0 denotes $F(x_0, y_0)$, u_1 denotes $F(x_0 \pm h, y_0 \pm k)$; by Taylor's series,

(1) $u_1 - u_0 = \dfrac{\partial F}{\partial x_0}(h) + \dfrac{\partial F}{\partial y_0}(k)$

$+ \frac{1}{2}\left\{ \dfrac{\partial^2 F}{\partial x_0^2} \cdot h^2 + 2 \dfrac{\partial^2 F}{\partial x_0 \partial y_0} \cdot hk + \dfrac{\partial^2 F}{\partial y_0^2} \cdot k^2 \right\}$

$+ \dfrac{1}{3!}\left\{ \dfrac{\partial^3 F}{\partial x_0^3} h^3 + 3 \dfrac{\partial^3 F}{\partial x_0^2 \partial y_0} \cdot h^2 k + 3 \dfrac{\partial^3 F}{\partial x_0 \partial y_0^2} \cdot hk^2 + \dfrac{\partial^3 F}{\partial y_0^2} \cdot k^3 \right\} + \cdots,$

where $\dfrac{\partial F}{\partial x_0}, \dfrac{\partial F}{\partial y_0}, \dfrac{\partial^2 F}{\partial x_0^2}, \cdots$ denote $\dfrac{\partial F}{\partial x}, \dfrac{\partial F}{\partial y}, \dfrac{\partial^2 F}{\partial x^2}, \cdots$ when $x = x_0$, $y = y_0$.

If h and k approach zero, $u_1 - u_0$ approaches the sum of the terms of the right-hand member of (1) which are of the lowest dimensions in h and k. If either or both $\dfrac{\partial F}{\partial x_0}, \dfrac{\partial F}{\partial y_0}$ are different from zero, $u_1 - u_0$ has different signs for different values of h and k, and u_0 is neither a maximum nor a minimum. If $\dfrac{\partial F}{\partial x_0} = 0$, $\dfrac{\partial F}{\partial y_0} = 0$, and $\dfrac{\partial^2 F}{\partial x_0^2}, \dfrac{\partial^2 F}{\partial x_0 \partial y_0}, \dfrac{\partial^2 F}{\partial y_0^2}$ are not each zero, u_0 is a maximum if $\dfrac{\partial^2 F}{\partial x_0^2} h^2 + 2 \dfrac{\partial^2 F}{\partial x_0 \partial y_0} hk + \dfrac{\partial^2 F}{\partial y_0^2} k^2$ is negative for all signs of h and k, a minimum if this expansion is positive for all signs of h and k.

Writing $\dfrac{\partial^2 F}{\partial x_0^2} \cdot h^2 + 2 \dfrac{\partial^2 F}{\partial x_0 \partial y_0} \cdot hk + \dfrac{\partial^2 F}{\partial y_0^2} \cdot k^2$

$\equiv \dfrac{\left(\dfrac{\partial^2 F}{\partial x_0^2} h + \dfrac{\partial^2 F}{\partial x_0 \partial y_0} k \right)^2 + \left\{ \dfrac{\partial^2 F}{\partial x_0^2} \cdot \dfrac{\partial^2 F}{\partial y_0^2} - \left(\dfrac{\partial^2 F}{\partial x_0^2 \partial y_0^2} \right)^2 \right\} k^2}{\dfrac{\partial^2 F}{\partial x_0^2}},$

it is seen that $u_1 - u_0$ is negative for all values of h and k, and consequently u_0 is a minimum, when $\dfrac{\partial^2 F}{\partial x_0^2}$ and $\dfrac{\partial^2 F}{\partial y_0^2}$ are both negative, and $\dfrac{\partial^2 F}{\partial x_0^2} \cdot \dfrac{\partial^2 F}{\partial y_0^2} > \left(\dfrac{\partial^2 F}{\partial x_0 \partial y_0}\right)^2$; $u_1 - u_0$ is positive, and u_0 a minimum, when $\dfrac{\partial^2 F}{\partial x_0^2}$ and $\dfrac{\partial^2 F}{\partial y_0^2}$ are both positive, and $\dfrac{\partial^2 F}{\partial x_0^2} \cdot \dfrac{\partial^2 F}{\partial y_0^2} > \left(\dfrac{\partial^2 F}{\partial x_0 \partial y_0}\right)^2$.

EXAMPLE. — Find the dimensions of the rectangular parallelopiped of maximum volume, sides parallel to the coordinate axes, that can be inscribed in the ellipsoid $\dfrac{x^2}{a^2} + \dfrac{y^2}{b^2} + \dfrac{z^2}{c^2} = 1$.

The volume of the parallelopiped is

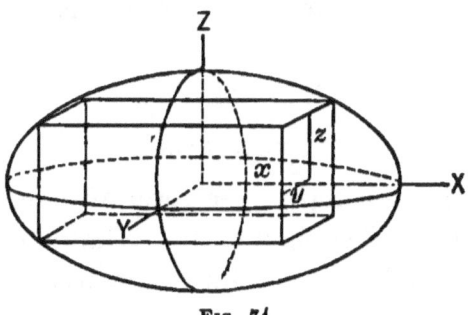

FIG. 74.

$$V = 8\,xyz = 8\,cxy\left(1 - \dfrac{x^2}{a^2} - \dfrac{y^2}{b^2}\right)^{\frac{1}{2}}.$$

If V is a maximum, $V_1 = x^2 y^2 - \dfrac{x^4 y^2}{a^2} - \dfrac{x^2 y^4}{b^2}$ is a maximum, and *vice versa*. Forming the partial derivatives of V_1,

$$\dfrac{\partial V_1}{\partial x} = 2\,xy^2 - \dfrac{4\,x^3 y^2}{a^2} - \dfrac{2\,xy^4}{b^2},$$

$$\dfrac{\partial V_1}{\partial y} = 2\,x^2 y - \dfrac{2\,x^4 y}{a^2} - \dfrac{4\,x^2 y^3}{b^2},$$

$$\dfrac{\partial^2 V_1}{\partial x^2} = 2\,y^3 - \dfrac{12\,x^2 y^2}{a^2} - \dfrac{2\,y^4}{b^2},$$

$$\frac{\partial^2 V_1}{\partial y^2} = 2\,x^2 - \frac{2\,x^4}{a^2} - \frac{12\,x^2 y^2}{b^2},$$

$$\frac{\partial^2 V_1}{\partial x\,\partial y} = 4\,xy - \frac{8\,x^3 y}{a^2} - \frac{8\,xy^3}{b^2}.$$

The conditions $\dfrac{\partial V_1}{\partial x}=0$, $\dfrac{\partial V_1}{\partial y}=0$ make $x=\dfrac{a}{\sqrt{3}}$, $y=\dfrac{b}{\sqrt{3}}$.
These values of x and y make

$$\frac{\partial^2 V_1}{\partial x^2} = -\frac{8\,b^2}{9},\quad \frac{\partial^2 V_1}{\partial y^2} = -\frac{8\,a^2}{9},\quad \frac{\partial^2 V_1}{\partial x\,dy} = -\frac{4\,ab}{9}.$$

Since $\dfrac{\partial^2 V_1}{\partial x^2}$ and $\dfrac{\partial^2 V_1}{\partial y^2}$ are both negative and $\dfrac{\partial^2 V_1}{\partial x^2}\cdot\dfrac{\partial^2 V_1}{\partial y^2} > \left(\dfrac{\partial^2 V_1}{\partial x\,\partial y}\right)^2$, V_1 is a maximum. Hence the dimensions of the maximum parallelopiped are $\dfrac{2\,a}{\sqrt{3}}$, $\dfrac{2\,b}{\sqrt{3}}$, $\dfrac{2\,c}{\sqrt{3}}$.

PROBLEMS

1. A box with open top in the form of a rectangular parallelopiped contains 108 cubic inches. What must be its dimensions to require the least material in construction?

2. Find the point (x, y, z) the sum of the squares of whose distances from (a_1, b_1, c_1), (a_2, b_2, c_2), (a_3, b_3, c_3) is a minimum.

3. The sum of the three dimensions of a rectangular parallelopiped is a. Find the dimensions when the volume is a maximum.

4. Find the dimensions of the cistern of maximum capacity that can be built out of 3000 square feet of sheet iron, the cistern being of the form of a rectangular parallelopiped and without lid.

5. The sum of the three dimensions of a rectangular parallelopiped is a. Find the dimensions when the surface is a maximum.

Art. 72. — Contact of Plane Curves

Plot the curves representing the equations $Y = F(X)$ and $y = f(x)$ to the same coordinate axes, and denote by Y_0 and y_0 the values of Y and y corresponding to $X = x_0$, $x = x_0$, by Y_1 and y_1 the values of Y and y corresponding to $X = x_0 \pm h$, $x = x_0 \pm h$. By Taylor's series,

$$Y_1 = Y_0 + \frac{dY_0}{dX_0}(\pm h) + \frac{d^2Y_0}{dX_0^2}\frac{(\pm h)^2}{2!} + \frac{d^3Y_0}{dX_0^3}\frac{(\pm h)^3}{3!} + \frac{d^4Y_0}{dX_0^4}\frac{(\pm h)^4}{4!} + \cdots,$$

$$y_1 = y_0 + \frac{dy_0}{dx_0}(\pm h) + \frac{d^2y_0}{dx_0^2}\frac{(\pm h)^2}{2!} + \frac{d^3y_0}{dx_0^3}\frac{(\pm h)^3}{3!} + \frac{d^4y_0}{dx_0^4}\frac{(\pm h)^4}{4!} + \cdots.$$

Hence

$$Y_1 - y_1 = (Y_0 - y_0) + \left(\frac{dY_0}{dX_0} - \frac{dy_0}{dx_0}\right)(\pm h) + \left(\frac{d^2Y_0}{dX_0^2} - \frac{d^2y_0}{dx_0^2}\right)\frac{(\pm h)^2}{2!}$$
$$+ \left(\frac{d^3Y_0}{dX_0^3} - \frac{d^3y_0}{dx_0^3}\right)\frac{(\pm h)^3}{3!} + \cdots.$$

The curves $Y = F(X)$, $y = f(x)$ have a common point if $Y_0 = y_0$ when $X_0 = x_0$. The difference between the ordinates

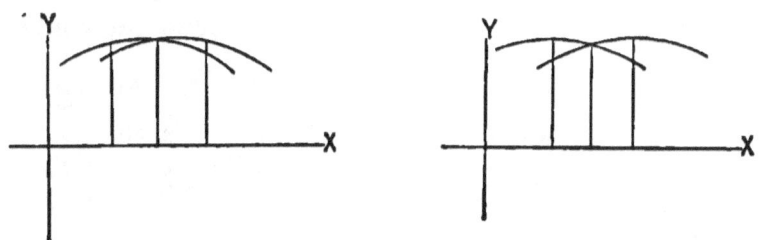

Fig. 75.

corresponding to $x_0 \pm h$ when h approaches zero is of the first degree in h if $\dfrac{dY_0}{dX_0} \neq \dfrac{dy_0}{dx_0}$; of the second degree in h and the curves are said to have contact of the first order, if

$$\frac{dY_0}{dX_0} = \frac{dy_0}{dx_0} \text{ and } \frac{d^2Y_0}{dX_0^2} \neq \frac{d^2y_0}{dx_0^2};$$

of the third degree in h and the curves are said to have contact of the second order if $\dfrac{dY_0}{dX_0} = \dfrac{dy_0}{dx_0}$, $\dfrac{d^2Y_0}{dX_0^2} = \dfrac{d^2y_0}{dx_0^2}$ and $\dfrac{d^3Y_0}{dX_0^3} \neq \dfrac{d^3y_0}{dx_0^3}$. In general, the difference between the ordinates is of the $(n+1)$th degree in h and the curves are said to have contact of the nth order when the first pair of corresponding derivatives not equal are of order $n+1$.

If the contact of the two curves $Y = F(X)$ and $y = f(x)$ at (x_0, y_0) is of an even order $2m$,

$$Y_1 - y_1 = \left(\frac{d^{2m+1}Y_0}{dX_0^{2m+1}} - \frac{d^{2m+1}y_0}{dx_0^{2m+1}}\right)\frac{(\pm h)^{2m+1}}{(2m+1)!}.$$

Hence $Y_1 - y_1$ changes sign with h and the curves intersect. If the contact is of an odd order, the curves do not intersect.

Suppose the equation $y = f(x)$ to be completely determined, while the equation $Y = F(X)$ involves arbitrary constants, that is, parameters. The condition necessary for the intersection of the curves represented by the equations $Y = F(X)$ and $y = f(x)$ when $X_0 = x_0$, namely $Y_0 = y_0$, determines one of the parameters of $Y = F(X)$; the conditions for contact of the first order when $X_0 = x_0$, namely $Y_0 = y_0$, $\dfrac{dY_0}{dX_0} = \dfrac{dy_0}{dx_0}$, determine two parameters of $Y = F(X)$; the conditions for contact of the second order $Y_0 = y_0$, $\dfrac{dY_0}{dX_0} = \dfrac{dy_0}{dx_0}$, $\dfrac{d^2Y_0}{dX_0^2} = \dfrac{d^2y_0}{dx_0^2}$ determine three parameters of $Y = F(X)$. In general, contact of the nth order determines $n+1$ parameters of $Y = F(X)$. It is also evident that the highest order of contact $Y = F(X)$ can in general have with another curve $y = f(x)$ is one less than the number of parameters of $Y = F(X)$.

EXAMPLE I. — Determine the order of contact of $4y = x^2 - 4$ and $x^2 + y^2 - 2y = 3$.

The common point is $(0, -1)$. For this point the first equation gives $\frac{dy}{dx} = 0$, $\frac{d^2y}{dx^2} = \frac{1}{2}$, $\frac{d^3y}{dx^3} = 0$, $\frac{d^4y}{dx^4} = 0$, $\cdots$, the second equation $\frac{dy}{dx} = 0$, $\frac{d^2y}{dx^2} = \frac{1}{2}$, $\frac{d^3y}{dx^3} = 0$, $\frac{d^4y}{dx^4} = -\frac{1}{8}$. Hence there is contact of the third order.

EXAMPLE II. — The equation $y = f(x)$ represents a fixed curve, $(X-m)^2 + (Y-n)^2 = R^2$ is the equation of any circle. The parameters m, n, R are to be so determined that the circle has contact of the second order with $y = f(x)$ at the point (x_0, y_0).

The problem requires that

(1) $\quad Y_0 = y_0, \quad \dfrac{dY_0}{dX_0} = \dfrac{dy_0}{dx_0}, \quad \dfrac{d^2Y_0}{dX_0^2} = \dfrac{d^2y_0}{dx_0^2}$

when $X_0 = x_0$. Differentiating the equation of the circle twice in succession,

$$(X-m)^2 + (Y-n)^2 = R^2, \quad (X-m) + (Y-n)\frac{dY}{dX} = 0,$$

$$1 + \left(\frac{dY}{dX}\right)^2 + (Y-n)\frac{d^2Y}{dX^2} = 0.$$

By the conditions (1), these equations become

$$(x_0-m)^2 + (y_0-n)^2 = R^2, \quad (x_0-m) + (y_0-n)\frac{dy_0}{dx_0} = 0,$$

$$1 + \frac{dy_0^2}{dx_0^2} + (y_0-n)\frac{d^2y_0}{dx_0^2} = 0.$$

Whence

$$R = \frac{\left(1+\dfrac{dy_0^2}{dx_0^2}\right)^{\frac{3}{2}}}{\dfrac{d^2y_0}{dx_0^2}}, \quad m = x_0 - \frac{\left(1+\dfrac{dy_0^2}{dx_0^2}\right)\dfrac{dy_0}{dx_0}}{\dfrac{d^2y_0}{dx_0^2}}, \quad n = y_0 + \frac{1+\dfrac{dy_0^2}{dx_0^2}}{\dfrac{d^2y_0}{dx_0^2}}.$$

APPLICATIONS OF TAYLOR'S SERIES   201

This circle is called the osculating circle at the point (x_0, y_0) of the curve $y = f(x)$, and by comparison with the results of Art. 44, this circle is seen to be identical with the circle of curvature.

PROBLEMS

Determine the order of contact of

1. $y^2 = 4x$ and $x^2 + y^2 = 4x$.
2. $x^2 y + y - x = 0$ and $y = 0$.
3. $\dfrac{x^2}{4} + y^2 = 1$ and $x^2 + y^2 + 6y - 7 = 0$.
4. Show that at a point of inflection the tangent $y = mx + n$ has contact of the second order with $y = f(x)$.
5. Show that at a point of maximum or minimum curvature of $y = f(x)$ the osculating circle has contact of the third order.

ART. 73. — SINGULAR POINTS OF PLANE CURVES

Let (x_0, y_0) be any point of the plane curve $F(x, y) = 0$, $(x_0 + h, y_0 + k)$ any other point. By Taylor's series

$$F(x_0 + h, y_0 + k) \equiv \frac{\partial F}{\partial x_0} \cdot h + \frac{\partial F}{\partial y_0} \cdot k$$

$$+ \tfrac{1}{2}\left(\frac{\partial^2 F}{\partial x_0^2} \cdot h^2 + 2 \frac{\partial^2 F}{\partial x_0 \, \partial y_0} \cdot hk + \frac{\partial^2 F}{\partial y_0^2} \cdot k^2\right) + \cdots = 0.$$

Denoting the point $(x_0 + h, y_0 + k)$ by (x, y), this series becomes

(1) $\dfrac{\partial F}{\partial x_0}(x - x_0) + \dfrac{\partial F}{\partial y_0}(y - y_0)$

$+ \tfrac{1}{2}\left\{ \dfrac{\partial^2 F}{\partial x_0^2}(x - x_0)^2 + 2 \dfrac{\partial^2 F}{\partial x_0 \, \partial y_0}(x - x_0)(y - y_0) + \dfrac{\partial^2 F}{\partial y_0^2}(y - y_0)^2 \right\}$

$+ \cdots = 0.$

If either or both $\dfrac{\partial F}{\partial x_0}$, $\dfrac{\partial F}{\partial y_0}$ differ from zero, when (x, y) indefinitely approaches (x_0, y_0), (1) approaches

$$\frac{\partial F}{\partial x_0}(x - x_0) + \frac{\partial F}{\partial y_0}(y - y_0) = 0,$$

the tangent to $F(x, y) = 0$ at (x_0, y_0).

If $\dfrac{\partial F}{\partial x_0} = 0$ and $\dfrac{\partial F}{\partial y_0} = 0$, (1) approaches

$$\frac{\partial^2 F}{\partial x_0^2}(x - x_0)^2 + 2\frac{\partial^2 F}{\partial x_0\, \partial y_0}(x - x_0)(y - y_0) + \frac{\partial^2 F}{\partial y_0^2}(y - y_0)^2 = 0.$$

This equation is homogeneous of the second degree in $(x-x_0)$ and $(y-y_0)$, and therefore represents two straight lines through (x_0, y_0). This means that at the point (x_0, y_0) of the curve

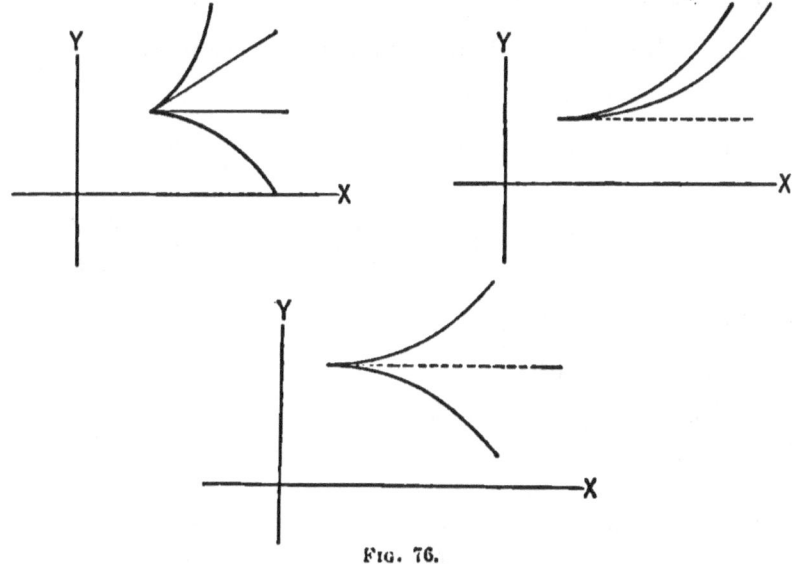

Fig. 76.

$F(x, y) = 0$ two tangents can be drawn to the curve. If the curve stops at (x_0, y_0) and the tangents are real and distinct, the curve is said to have a salient point at (x_0, y_0); if the tangents are real and coincident, the curve is said to have a cusp at (x_0, y_0), of the first species when the two branches of the curve lie on different sides of the common tangent, of the second species when both curves lie on the same side of the common tangent.

If the curve does not stop at (x_0, y_0) and the tangents are real and distinct, (x_0, y_0) is a point where the curve crosses itself, called a node.

If the tangents are real and coincident, two branches of the curve are tangent to each other at (x_0, y_0), and (x_0, y_0) is called a tac-node.

If the tangents at (x_0, y_0) are imaginary, (x_0, y_0) is an isolated point, called a conjugate point of the curve.

An ordinary or regular point of a curve is a point (x_0, y_0) for which $\dfrac{\partial F}{\partial x}$ and $\dfrac{\partial F}{\partial y}$ are not both zero; all other points are singular. The points of inflection of a curve, where $\dfrac{d^2y}{dx^2}=0$, are also classed as singular points.

EXAMPLE I. — Examine $x^3 - 3xy + y^3 = 0$ for singular points. Here

$$\frac{\partial F}{\partial x}=3x^2-3y,\ \frac{\partial F}{\partial y}=-3x+3y^2,\ \frac{\partial^2 F}{\partial x^2}=6x,\ \frac{\partial^2 F}{\partial y^2}=6y,\ \frac{\partial^2 F}{\partial x \partial y}=-3.$$

The conditions $\dfrac{\partial F}{\partial x}=0$, $\dfrac{\partial F}{\partial y}=0$ determine the singular point $(0, 0)$. For the point $(0, 0)$,

$$\frac{\partial^2 F}{\partial x^2}=0,\ \frac{\partial^2 F}{\partial y^2}=0,\ \frac{\partial^2 F}{dx\,dy}=-3.$$

Hence the two tangents at $(0, 0)$ are $x = 0$ and $y = 0$, the coordinate axes. Plotting the curve in the neighborhood of $(0, 0)$, it is seen that the origin is a node.

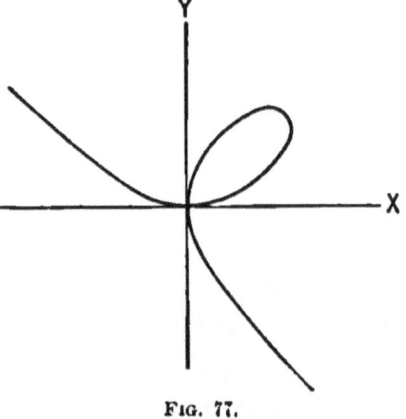

FIG. 77.

EXAMPLE II. — Examine $y^2 - x^3 + 2x^2 = 0$ for singular points. Here
$$\frac{\partial F}{\partial x} = -3x^2 + 4x, \quad \frac{\partial F}{\partial y} = 2y, \quad \frac{\partial^2 F}{\partial x^2} = -6x + 4, \quad \frac{\partial^2 F}{\partial y^2} = 2, \quad \frac{\partial^2 F}{\partial x \partial y} = 0.$$
The conditions $\frac{\partial F}{\partial x} = 0$, $\frac{\partial F}{\partial y} = 0$ determine the singular point $(0, 0)$. For this point $\frac{\partial^2 F}{\partial x^2} = 4$, $\frac{\partial^2 F}{\partial y^2} = 2$, $\frac{\partial^2 F}{\partial x \partial y} = 0$. Hence the tangents at $(0, 0)$, represented by $4x^2 + 2y^2 = 0$, are $y = \pm \sqrt{-2} \cdot x$, and the point $(0, 0)$ is an isolated point of the curve.

EXAMPLE III. — Examine $y^2 = x^3$ for singular points.
Here $\frac{\partial F}{\partial x} = -3x^2$, $\frac{\partial F}{\partial y} = 2y$, $\frac{\partial^2 F}{\partial x^2} = -6x$, $\frac{\partial^2 F}{\partial y^2} = 2$, $\frac{\partial^2 F}{\partial x \partial y} = 0$.
The conditions $\frac{\partial F}{\partial x} = 0$, $\frac{\partial F}{\partial y} = 0$ determine the singular point $(0, 0)$. For this point $\frac{\partial^2 F}{\partial x^2} = 0$, $\frac{\partial^2 F}{\partial y^2} = 2$, $\frac{\partial^2 F}{\partial x \partial y} = 0$. Hence the two tangents at $(0, 0)$, represented by $2y^2 = 0$, coincide with the X-axis. Plotting the curve in the neighborhood of $(0, 0)$, it is seen that the origin is a cusp of the first species.

PROBLEMS

Examine for singular points,

1. $x^3 + 3x^2 - y^2 + 3x + 4y - 1 = 0$.
2. $(y - x^2)^2 = x^5$.
3. $y^2 = \dfrac{x^3}{x - 2}$.
4. $y^2 = x^2 + 2x^3$.
5. $y^2 = x^4 + x^5$.
6. $(x^2 + y^2)^2 = a^2(x^2 - y^2)$.
7. $(xy + 1)^2 + (x - 1)^3(x - 2) = 0$.
8. $y^2 = x^4 - x^6$.

CHAPTER XII

ORDINARY DIFFERENTIAL EQUATIONS OF FIRST ORDER

Art. 74. — Formation of Differential Equations

An equation containing ordinary derivatives

$$F\left(x,\ y,\ \frac{dy}{dx},\ \frac{d^2y}{dx^2}\right) = 0$$

is called an ordinary differential equation.

An equation containing partial derivatives

$$F\left(x,\ y,\ z,\ \frac{\partial z}{\partial x},\ \frac{\partial z}{\partial y},\ \frac{\partial^2 z}{\partial x^2},\ \frac{\partial^2 z}{\partial y^2}\right) = 0$$

is called a partial differential equation.

The order of a differential equation is the order of the highest order derivative occurring in the equation.

The degree of a differential equation is the greatest exponent of the highest order derivative when the exponents of all derivatives in the equation are positive integers.

Example I. — The equation $(x-c)^2 + y^2 = \tfrac{1}{4}c^2$ represents all circles with center in the X-axis and with radius $\tfrac{1}{2}$ the abscissa of the center. Differentiating this equation with respect to x, $x - c + y\dfrac{dy}{dx} = 0$. Eliminating c from this result and the given equation, $3\,y^2\dfrac{dy^2}{dx^2} - 2\,xy\dfrac{dy}{dx} + 4\,y^2 - x^2 = 0$, the differential equation of the given system of circles. Solving

the differential equation for $\dfrac{dy}{dx}$, $\dfrac{dy}{dx} = \dfrac{2xy \pm \sqrt{16 x^2 y^2 - 48 y^4}}{6 y^2}$.
Hence for every point (x, y) of the plane where $16 x^2 y^2 - 48 y^4 > 0$, the differential equation determines two unequal values for $\dfrac{dy}{dx}$; where $16 x^2 y^2 - 48 y^4 = 0$, the two values of $\dfrac{dy}{dx}$ are equal; where $16 x^2 y^2 - 48 y^4 < 0$, the two values of $\dfrac{dy}{dx}$ are imaginary. Geometrically these results mean that through points (x, y) for which $-\dfrac{x}{\sqrt{3}} < y < +\dfrac{x}{\sqrt{3}}$ two circles of the system pass and their tangents at (x, y) have different directions; through

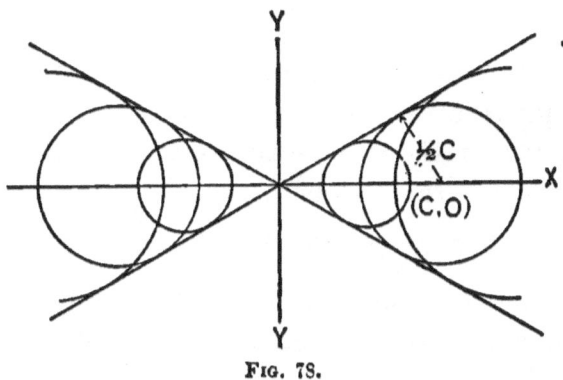

Fig. 78.

points (x, y) for which $y = \pm \dfrac{x}{\sqrt{3}}$ pass two circles which have a common direction at this point; and through points (x, y) for which $y^2 > \dfrac{x^3}{3}$ no circles of the system pass.

If x, y, $\dfrac{dy}{dx}$ satisfy the differential equation, $\left(x, y, \dfrac{dy}{dx}\right)$ denotes a point in the circumference of one of the circles of the system and moving along the circumference. If $\left(x, y, \dfrac{dy}{dx}\right)$ moves along in obedience to the differential equation, changing its direction continuously, it describes the circumference on which it started.

If, however, $\left(x, y, \dfrac{dy}{dx}\right)$ denotes a point in either of the lines $y = \pm \dfrac{x}{\sqrt{3}}$, it may move along in obedience to the differential equation without a discontinuous change of direction, and describe the straight lines $y = \pm \dfrac{x}{\sqrt{3}}$. These straight lines are the envelopes of the system of circles $(x-c)^2 + y^2 = \tfrac{1}{4} c^2$.

The equation $(x-c)^2 + y^2 = \tfrac{1}{4} c^2$ is called the general solution of the differential equation

$$3 y^2 \dfrac{dy^2}{dx^2} - 2 xy \dfrac{dy}{dx} + 4 y^2 - x^2 = 0.$$

The solution obtained by assigning to the arbitrary constant in the general solution some particular value is called a particular solution of the differential equation and represents a particular circle of the system.

The solution $y = \pm \dfrac{x}{\sqrt{3}}$, which cannot be obtained from the general solution by assigning a particular value to the arbitrary constant, is called a singular solution of the differential equation. If the differential equation is written $F(x, y, p) = 0$, where $p = \dfrac{dy}{dx}$, the preceding analysis shows that the singular solution is the p-envelope of the equation.

EXAMPLE II. — Form the differential equation of the system of circles $x^2 + y^2 - 2ax - 2by + c = 0$.

Differentiating three terms in succession,

(1) $\quad x + y\dfrac{dy}{dx} - a - b\dfrac{dy}{dx} = 0,$

(2) $\quad 1 + \left(\dfrac{dy}{dx}\right)^2 + y\dfrac{d^2y}{dx^2} - b\dfrac{d^2y}{dx^2} = 0 \quad \text{or} \quad \dfrac{1 + \left(\dfrac{dy}{dx}\right)^2 + y\dfrac{d^2y}{dx^2}}{\dfrac{d^2y}{dx^2}} = b,$

$$(3) \quad 3\frac{dy}{dx}\left(\frac{d^2y}{dx^2}\right)^2 - \frac{d^3y}{dx^3} - \left(\frac{dy}{dx}\right)^2\frac{d^3y}{dx^3} = 0.$$

Observe that in Examples I. and II. the order of the differential equation is the same as the number of arbitrary constants in the general solution. This is always the case.

The solution of a differential equation is also called the primitive of the equation. The differential equation is obtained from its primitive either directly by differentiation or by the elimination of constants from the primitive and its derivatives. The process of finding the primitive of a differential equation is called solving the equation.*

PROBLEMS

Form the differential equations of the following primitives,

1. $y = cx + c - c^3$.
2. $y = c_1 x^3 + \dfrac{c_2}{x}$.
3. $(y + c)^2 = 4\, ax$.
4. $y = c_1 e^{ax} + c_2 e^{-ax}$.
5. $y = c_1 \cos(ax + c_2)$.
6. $y = c_1 e^{2x} + c_2 e^{-3x} + c_3 e^x$.

7. Form the differential equation of the system of straight lines $y = mx + n$.

8. Form the differential equation of the system of circles concentric at the origin.

9. Form the differential equation of the system of parabolas $y^2 = 2\,px$.

---

* The mathematical expression of every physical law leads to a differential equation. For example, the relation between current i and time t in a circuit whose constants are R, L, C is expressed by the second order differential equation $\dfrac{d^2i}{dt^2} + \dfrac{R}{L}\dfrac{di}{dt} + \dfrac{i}{LC} = \dfrac{1}{L}f'(t)$, where $f(t)$ is the electromotive force expressed as a function of time.

10. Form the differential equation of the system of ellipses $\dfrac{x^2}{a^2}+\dfrac{y^2}{b^2}=1$.

11. Form the differential equation of the system of tangents $y = sx \pm \sqrt{1+s^2}$ to $x^2+y^2=1$ and find the singular solution of the differential equation.

12. A point (x, y) generates a curve. Write the differential equation which expresses the fact that the angle the line from the origin to (x, y) makes with the X-axis is the supplement of the angle the direction of the point (x, y) makes with the X-axis.

Art. 75. — Solution of First Order Differential Equations of First Degree

First order differential equations which are readily solved occur in the following standard forms:

Standard I. — $XY\dfrac{dy}{dx} + X_1 Y_1 = 0$, where X, X_1 are functions of x only; Y, Y_1 functions of y only. Dividing the equation by XY_1, $\dfrac{Y}{Y_1}\dfrac{dy}{dx} + \dfrac{X_1}{X} = 0$, the variables are separated and each term may be integrated.

Example. — Solve $(x^2 - yx^2)\dfrac{dy}{dx} + y^2 + xy^2 = 0$. Writing the equation $\dfrac{dy}{y^2(1-y)} + \dfrac{1+x}{x}dx = 0$, and integrating term by term, $\log\dfrac{x}{y} - \dfrac{y+x}{xy} = c$.

Standard II. — $\dfrac{dy}{dx} = \dfrac{f_1(x, y)}{f_2(x, y)}$, where $f_1(x, y)$ and $f_2(x, y)$ are homogeneous functions of the same degree. Now, a homogeneous function of degree n,

$$ax^n + bx^{n-1}y + cx^{n-2}y^2 + \cdots hx^2y^{n-2} + kxy^{n-1} + ly^n,$$

may be written

$$x^n\left(a + b\frac{y}{x} + c\frac{y^2}{x^2} + \cdots h\frac{y^{n-2}}{x^{n-2}} + k\frac{y^{n-1}}{x^{n-1}} + l\frac{y^n}{x^n}\right).$$

Hence, $\dfrac{dy}{dx} = \dfrac{f_1(x, y)}{f_2(x, y)} = F\left(\dfrac{y}{x}\right)$. Substituting $\dfrac{y}{x} = z$, $\dfrac{dy}{dx} = z + x\dfrac{dz}{dx}$, the given equation becomes $z + x\dfrac{dz}{dx} = F(z)$, whence $\dfrac{dx}{x} = \dfrac{dz}{F(z) - z}$, where the variables are separated.

EXAMPLE. — Solve $y^2 + x^2\dfrac{dy}{dx} = xy\dfrac{dy}{dx}$.

Here $\dfrac{dy}{dx} = \dfrac{y^2}{xy - x^2} = \dfrac{\frac{y^2}{x^2}}{\frac{y}{x} - 1}$. Substituting $\dfrac{y}{x} = z$, $\dfrac{dy}{dx} = z + x\dfrac{dz}{dx}$, the given equation becomes

$$z + x\frac{dz}{dx} = \frac{z^2}{z - 1}, \text{ whence } \frac{dx}{x} = \left(1 - \frac{1}{z}\right)dz.$$

Integrating, $\log x = z - \log z + \log c$, or $\dfrac{xz}{c} = e^z$. Substituting $z = \dfrac{y}{x}$, $y = c \cdot e^{\frac{y}{x}}$.

STANDARD III. — $(ax + by + c)\dfrac{dy}{dx} + Ax + By + C = 0$, where a, b, c, A, B, C are constants. Substituting (1) $x = x_0 + x_1$, $y = y_0 + y_1$, whence $\dfrac{dy}{dx} = \dfrac{dy_1}{dx_1}$,

$$\{(ax_0 + by_0 + c) + (ax_1 + by_1)\}\frac{dy_1}{dx_1} + (Ax_0 + By_0 + C) + (Ax_1 + By_1) = 0.$$

Determining x_0, y_0 so that $ax_0 + by_0 + c = 0$, $Ax_0 + By_0 + C = 0$, whence $x_0 = \dfrac{Bc - bC}{Ab - aB}$, $y_0 = \dfrac{Ca - cA}{Ab - aB}$, the equation becomes $(ax_1 + by_1)\dfrac{dy_1}{dx_1} + Ax_1 + By_1 = 0$, which is homogeneous of the first degree.

The substitution (1) is impossible when $Ab - aB = 0$. In this case, writing $\frac{A}{a} = \frac{B}{b} = m$, the given equation becomes $(ax + by + c)\frac{dy}{dx} + m(ax + by) + C = 0$, where the variables are separated by the substitution $ax + by = z$.

EXAMPLE. — Solve $(3y - 7x + 7) + (7y - 3x + 3)\frac{dy}{dx} = 0$.

Substituting $x = x_0 + x_1$, $y = y_0 + y_1$, the given equation becomes

$$\{(7y_0 - 3x_0 + 3) + (7y_1 - 3x_1)\}\frac{dy_1}{dx_1}$$
$$+ \{(3y_0 - 7x_0 + 7) + (3y_1 - 7x_1)\} = 0.$$

Writing $7y_0 - 3x_0 + 3 = 0$, $3y_0 - 7x_0 + 7 = 0$, whence $x_0 = 1$, $y_0 = 0$, there results $(7y_1 - 3x_1)\frac{dy_1}{dx_1} + 3y_1 - 7x_1 = 0$. This equation may be written $\frac{dy_1}{dx_1} = \frac{7 - 3\frac{y_1}{x_1}}{7\frac{y_1}{x_1} - 3}$, which, by the substitution $\frac{y_1}{x_1} = z$, $\frac{dy_1}{dx_1} = z + x_1\frac{dz}{dx_1}$, becomes $\frac{dx_1}{x_1} = \frac{7z - 3}{7(1 - z^2)}dz$. Integrating by partial fractions,

$$\log x_1 = -\tfrac{5}{7}\log(1 + z) - \tfrac{2}{7}\log(1 - z) + \log c.$$

Substituting $z = \frac{y_1}{x_1}$, there results, finally,

$$(x + y - 1)^5(x - y + 1)^2 = C.$$

STANDARD IV. — $\frac{dy}{dx} + X_1 y = X_2$, where X_1 and X_2 are functions of x only. Since this equation is of the first degree in y and its derivatives, it is called the linear equation of the first order.

Consider the equation $\frac{dy}{dx} + X_1 y = 0$. Writing this in the form $\dfrac{dy}{y} = -X_1\, dx$ and integrating, $y = c \cdot e^{-\int X_1 dx}$ or $y \cdot e^{\int X_1 dx} = c$.

Differentiating this result, $e^{\int X_1 dx}\left(\dfrac{dy}{dx} + X_1 y\right) = 0$.

To solve $\dfrac{dy}{dx} + X_1 y = X_2$, multiply both sides of the equation by $e^{\int X_1 dx}$, which gives $e^{\int X_1 dx}\left(\dfrac{dy}{dx} + X_1 y\right) = e^{\int X_1 dx} X_2$. Integrating, $y \cdot e^{\int X_1 dx} = \int e^{\int X_1 dx} X_2\, dx + C$, whence

$$y = e^{-\int X_1 dx}\left[\int e^{\int X_1 dx} X_2\, dx + C\right].*$$

The equation $\dfrac{dy}{dx} + X_1 y = X_2 y^n$ is reduced to the linear form by dividing by y^n. This gives $y^{-n}\dfrac{dy}{dx} + X_1 y^{1-n} = X_2$. Now

$$y^{-n}\dfrac{dy}{dx} = \dfrac{d}{dx}\int y^{-n}\dfrac{dy}{dx} = \dfrac{1}{1-n}\dfrac{d}{dx} y^{1-n}.$$

Hence the given equation becomes

$$\dfrac{d}{dx} y^{1-n} + (1-n) X_1 \cdot y^{1-n} = (1-n) X_2,$$

which is linear if y^{1-n} is considered the dependent variable.

Any equation of the form $f'(y)\dfrac{dy}{dx} + X_1 f(y) = X_2$ becomes linear by the substitution $z = f(y)$, $\dfrac{dz}{dx} = f'(y)\dfrac{dy}{dx}$.

EXAMPLE I. — Solve $(1 + x^2)\dfrac{dy}{dx} - xy = a$.

Writing the equation $\dfrac{dy}{dx} - \dfrac{x}{1+x^2} y = \dfrac{a}{1+x^2}$, it is seen to be linear with $X_1 = \dfrac{-x}{1+x^2}$, $X_2 = \dfrac{a}{1+x^2}$. Hence

$\int X_1\, dx = \log(1 + x^2)^{-\frac{1}{2}}$ and $e^{\int X_1 dx} = (1 + x^2)^{-\frac{1}{2}}$.

* Leibnitz seems to have been the first to obtain this formula.

Substituting in the formula,

$$y = (1+x^2)^{\frac{1}{2}} \left\{ \int \frac{a\,dx}{(1+x^2)^{\frac{3}{2}}} + C \right\}. \text{ Now } \int \frac{a\,dx}{(1+x^2)^{\frac{3}{2}}} = \frac{ax}{(1+x^2)^{\frac{1}{2}}},$$

found by substituting $x = \tan\theta$. Finally $y = ax + C(1+x^2)^{\frac{1}{2}}$.

EXAMPLE II. — Solve $\dfrac{dy}{dx} + y = xy^3$.

Dividing by y^3, $y^{-3}\dfrac{dy}{dx} + y^{-2} = x$. Writing $y^{-3}\dfrac{dy}{dx} = \dfrac{d}{dx}(-\tfrac{1}{2}y^{-2})$, the equation becomes $\dfrac{d}{dx}(-y^{-2}) - 2y^{-2} = -2x$ and

$$y^{-2} = e^{\int 2\,dx} \int e^{-\int 2\,dx}(-2x)\,dx = e^{2x} \int e^{-2x}(-2x)\,dx.$$

Integrating by parts,

$$y^{-2} = e^{2x}(xe^{-2x} + \tfrac{1}{2}e^{-2x} + C) = x + \tfrac{1}{2} + Ce^{2x}.$$

EXAMPLE III. — Solve $3y^2\dfrac{dy}{dx} - ay^3 = x+1$.

Writing this equation $\dfrac{d}{dx}(y^3) - ay^3 = x+1$,

$$y^3 = e^{ax}\left[\int e^{-ax}(x+1)\,dx + C\right] = Ce^{ax} - \frac{x+1}{a} - \frac{1}{a^2}.$$

STANDARD V. — *Exact equations.* A differential equation is said to be exact when it can be obtained directly by the differentiation of its primitive. By Art. 33 the equation $P\,dx + Q\,dx = 0$, where P and Q are functions of x and y, is exact if $\dfrac{\partial P}{\partial y} = \dfrac{\partial Q}{\partial x}$. The primitive is found by the method of Art. 33.

EXAMPLE. — Solve $x(x^2 + 3y^2)\,dx + y(y^2 + 3x^2)\,dy = 0$.

$\dfrac{\partial}{\partial y}(x^3 + 3xy^2) = 6xy = \dfrac{\partial}{\partial x}(y^3 + 3x^2y)$, hence the equation is exact. Considering y constant,

$$\int (x^3 + 3xy^2)\, dx = \tfrac{1}{4} x^4 + \tfrac{3}{2} x^2 y^2 + f_1(y).$$

Differentiating this result partially with respect to y, and equating to coefficient of dy in given equation,

$$3 x^2 y + \frac{d}{dy} f_1(y) = y^3 + 3 x^2 y,$$

whence $\quad \dfrac{d}{dy} f_1(y) = y^3 \ \text{ and }\ f_1(y) = \tfrac{1}{4} y^4 + C.$

The primitive of the given equation is $\tfrac{1}{4} y^4 + \tfrac{3}{2} x^2 y^2 + \tfrac{1}{4} x^4 + C = 0$.

STANDARD VI. — *Integrating factor.* If the equation $P\, dx + Q\, dy = 0$ is not exact, a factor μ may be found for which the equation $\mu P\, dx + \mu Q\, dy = 0$ is exact. μ must satisfy the equation $\dfrac{\partial}{\partial y}(\mu P) = \dfrac{\partial}{\partial x}(\mu Q)$.

If μ is a function of x only, this equation becomes

$$\mu \frac{\partial P}{\partial y} = \mu \frac{\partial Q}{\partial x} + Q \frac{\partial \mu}{\partial x}, \text{ whence } \frac{1}{\mu} \frac{d\mu}{dx} = \frac{1}{Q}\left(\frac{\partial P}{\partial y} - \frac{\partial Q}{\partial x}\right)$$

and $\quad \mu = e^{\int \frac{1}{Q}\left(\frac{\delta P}{\delta y} - \frac{\delta Q}{\delta x}\right) dx}.$

If the equation $P\, dx + Q\, dy = 0$ is homogeneous of degree m and the integrating factor μ homogeneous of degree n, the equation $\mu P\, dx + \mu Q\, dy = 0$ is exact and homogeneous of degree $m + n$. Hence, by Problem 12, Art. 33,

$$\mu P x + \mu Q y = (m + n + 1) C.$$

Since C is an arbitrary constant, $(m + n + 1) C$ may be taken equal to unity and the integrating factor $\mu = \dfrac{1}{Px + Qy}$.

Since $\tfrac{1}{k} d(x^m y^n)^k = x^{km-1} y^{kn-1}(my\, dx + nx\, dy)$, the differential expression $x^\alpha y^\beta (my\, dx + nx\, dy)$ is rendered exact by the factor $x^{km-\alpha-1} y^{kn-\beta-1}$, where k is any number whatever.

EXAMPLE I. — Find the integrating factor of the linear equation $\frac{dy}{dx} + X_1 y = X_2$.

Supposing the factor to be a function of x only,

$$P = (X_1 y - X_2), \quad Q = 1, \quad \text{and} \quad \mu = e^{\int X_1 dx}.$$

EXAMPLE II. — Find the integrating factor of the homogeneous equation $(xy + y^2)dx - (x^2 - xy)dy = 0$.

The factor is $\mu = \dfrac{1}{x^2 y + xy^2 - x^2 y + xy^2} = \dfrac{1}{2\,xy^2}$.

Hence
$$\frac{dx}{2\,x} + \frac{dy}{2\,y} + \frac{dx}{2\,y} - \frac{x\,dy}{2\,y^2} = 0$$

is exact. Writing this equation $\dfrac{dx}{x} + \dfrac{dy}{y} + \dfrac{y\,dx - x\,dy}{y^2} = 0$ and integrating, $\log(xy) + \dfrac{x}{y} = C$.

EXAMPLE III. — Solve $(2\,x^2 y^2 + y)dx - (x^3 y - 3\,x)dy = 0$.

Break up the equation into two parts of the form

$$x^\alpha y^\beta (my\,dx + nx\,dy), \quad x^2 y(2\,y\,dx - x\,dy) + (y\,dx + 3\,x\,dy) = 0.$$

$x^{2k-3} y^{-k-2}$ is an integrating factor of the first part, $x^{k_1 - 1} y^{3k_1 - 1}$ an integrating factor of the second part. These factors are the same when $2k - 3 = k_1 - 1$, $-k - 2 = 3k_1 - 1$, whence $k_1 = -\frac{4}{7}$ and the common integrating factor of both parts of the equation is $x^{-\frac{11}{7}} y^{-\frac{19}{7}}$. Multiplying the equation by this factor and integrating, $\frac{7}{5} x^{\frac{10}{7}} y^{-\frac{5}{7}} - \frac{7}{4} x^{-\frac{4}{7}} y^{-\frac{12}{7}} = C$, whence

$$4\,x^2 y = 5 + C x^{\frac{4}{7}} y^{\frac{12}{7}}.$$

STANDARD VII. — The equation $f_1(xy) y\,dx + f_2(xy) x\,dy = 0$ may be solved by the substitution $xy = v$, whence

$$dy = \frac{x\,dv - v\,dx}{x^2}.$$

The equation becomes $f_1(v)\dfrac{v}{x}dx + f_2(v)\dfrac{xdv - v\,dx}{x} = 0$, reducing to $\dfrac{dx}{x} = \dfrac{f_2(v)dv}{v\{f_2(v) - f_1(v)\}}$, where the variables are separated.

EXAMPLE. — Solve $(x^2y^2 + xy)y\,dx + (x^2y^2 - 1)x\,dy = 0$.

Substituting $xy = v$, $dy = \dfrac{xdv - v\,dx}{x^2}$, the equation becomes

$(v^2 + v)\dfrac{v}{x}dx + (v^2 - 1)\dfrac{xdv - v\,dx}{x} = 0$, which reduces to $\dfrac{dx}{x} = \dfrac{dv}{v}$.

Integrating, $\log x = \log v + \log c$, $x = e^{cv}$ and finally $x = e^{cxy}$.

In attempting to solve a differential equation, determine what standard applies and proceed by the method of that standard.

PROBLEMS

Solve,

1. $x^2\dfrac{dy}{dx} - y^2 - xy = 0$.

2. $\dfrac{dy}{dx} - \dfrac{a}{1-x}y = b$.

3. $(1-x)^2 y\,dx + (1+y)x^2\,dy = 0$.

4. $\dfrac{dy}{dx} + y = xy^3$.

9. $(y - 3x + 3)\dfrac{dy}{dx} = 2y - x - 4$.

5. $x\dfrac{dy}{dx} - y = \sqrt{x^2 - y^2}$.

10. $\dfrac{dy}{dx} = \sqrt{\dfrac{1 - y^2}{1 - x^2}}$.

6. $(1 + y^2) - x^{\frac{1}{2}}\dfrac{dy}{dx} = 0$.

11. $(x + y)^2\dfrac{dy}{dx} = a^2$.

7. $\dfrac{dy}{dx} + y = e^{-x}$.

12. $x\dfrac{dy}{dx} + \dfrac{y^2}{x} = y$.

8. $x^2y\,dx - (x^3 + y^3)\,dy = 0$.

13. $x\dfrac{dy}{dx} - y = \sqrt{x^2 + y^2}$.

14. $(x^2 + 1)\dfrac{dy}{dx} + 2xy = 4x^2$.

15. $\sec^2 x \cdot \tan y \cdot dx + \sec^2 y \cdot \tan x \cdot dy = 0$.

16. $(y - x)\dfrac{dy}{dx} + y = 0$.

17. $\dfrac{dy}{dx} = x^3 y^3 - xy.$ 18. $\dfrac{dy}{dx} = ay^2 x.$

19. $(\sqrt{xy} - 1) x\, dy - (\sqrt{xy} + 1) y\, dx = 0.$

20. $(xy - x^2) \dfrac{dy}{dx} = y^2.$

21. $xy\, dy - y^2\, dx = (x + y)^2 e^{-\tfrac{y}{x}} dx.$

22. $(x - y)^2 + 2xy \dfrac{dy}{dx} = 0.$

23. $(x^2 + 2xy - y^2)\, dx = (x^2 - 2xy - y^2)\, dy.$

24. $\dfrac{dx}{x} + \dfrac{dy}{y} + 2\left(\dfrac{dx}{y} - \dfrac{dy}{x}\right) = 0.$ 25. $\dfrac{dy}{dx} = \dfrac{x^2 + y^2}{2xy}.$

26. $y\,(xy + 2 x^2 y^2)\, dx + x\,(xy - x^2 y^2)\, dy = 0.$

27. $\dfrac{dy}{dx} \cos x + y \sin x = 1.$

28. $(x^3 + 3 xy^2)\, dx + (y^3 + 3 x^2 y)\, dy = 0.$

29. $(y^3 - 2 yx^2)\, dx + (2 xy^2 - x^3)\, dy = 0.$

30. $(x + y) \dfrac{dy}{dx} + x - y = 0.$ 31. $\dfrac{dy}{dx} + y \cos x = \sin(2x).$

32. $\dfrac{dy}{dx} = \dfrac{1 + y + y^2}{1 + x + x^2}.$

33. $x\,(x^2 - 3 y^2) + y\,(3 x^2 - y^2) \dfrac{dy}{dx} = 0.$

34. $\dfrac{dy}{dx} + e^x y = e^x y^2.$ 36. $(1 + x^2) \dfrac{dy}{dx} + xy - \dfrac{1}{x} = 0.$

35. $\dfrac{dy}{dx} = \dfrac{7 y + x + 2}{3 x + 5 y + 6}.$ 37. $y \dfrac{dy}{dx} + by^2 = a \cos x.$

38. $(1 + xy) y\, dx + (1 - xy) x\, dy = 0.$

39. Determine the curve whose subtangent is constant.

40. Determine the curve whose subnormal is constant.

41. Determine the curve whose subtangent at any point equals the sum of the coordinates of that point.

42. The radius vector cuts a curve under a constant angle. Find the equation of the curve.

43. Find the system of curves which intersect all parabolas $y^2 = 2px$ at right angles.

Through every point (x, y) of the plane there passes one parabola, whose direction at this point is $\dfrac{dy}{dx} = \dfrac{p}{y} = \dfrac{y}{2x}$. For the curve which cuts this parabola at (x, y) at right angles, $\dfrac{dy}{dx} = -\dfrac{2x}{y}$. Integrating, $\dfrac{y^2}{2c^2} + \dfrac{x^2}{c^2} = 1$, a system of ellipses.

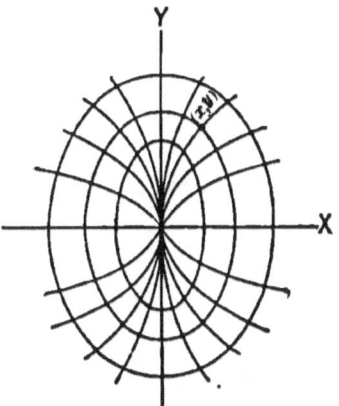

Fig. 79.

44. Find the system of curves intersecting the hyperbolas $xy = a^2$ at right angles.

Art. 76. — Equations of First Order and Higher Degrees

Case I. — Suppose the equation of the nth degree

$$\frac{dy^n}{dx^n} + p_1\frac{dy^{n-1}}{dx^{n-1}} + p_2\frac{dy^{n-2}}{dx^{n-2}} + \cdots + p_{n-2}\frac{dy^2}{dx^2} + p_{n-1}\frac{dy}{dx} + p_n = 0$$

to be resolvable into n factors,

$$\left(\frac{dy}{dx} - q_1\right)\left(\frac{dy}{dx} - q_2\right)\left(\frac{dy}{dx} - q_3\right)\cdots\left(\frac{dy}{dx} - q_n\right) = 0.$$

The given equation is satisfied only when one of these factors vanishes. Representing by

$$f_1(x, y, c_1) = 0, \; f_2(x, y, c_2) = 0, \; \cdots f_n(x, y, c_n) = 0$$

the primitives of the n first degree differential equations obtained by equating to zero the n factors, the product of these n primitives

$$f_1(x, y, c_1) f_2(x, y, c_2) f_3(x, y, c_3) \cdots f_n(x, y, c_n) = 0$$

includes all the partial solutions of the given equation. Since in each partial solution the constant may have all values from $+\infty$ through 0 to $-\infty$, all possible values of the partial solutions are included in the product

$$f_1(x, y, c_1) f_2(x, y, c) \cdots f_n(x, y, c) = 0,$$

where c is an arbitrary constant. This last product is the general solution of the given equation.

EXAMPLE I. — Solve $\dfrac{dy^2}{dx^2} - ax = 0$.

Write the equation $\left(\dfrac{dy}{dx} + a^{\frac{1}{2}} x^{\frac{1}{2}}\right)\left(\dfrac{dy}{dx} - a^{\frac{1}{2}} x^{\frac{1}{2}}\right) = 0$, and solve the equations $\dfrac{dy}{dx} + a^{\frac{1}{2}} x^{\frac{1}{2}} = 0$, $\dfrac{dy}{dx} - a^{\frac{1}{2}} x^{\frac{1}{2}} = 0$. The product of these solutions, $(y + \frac{2}{3} a^{\frac{1}{2}} x^{\frac{3}{2}} + c)(y - \frac{2}{3} a^{\frac{1}{2}} x^{\frac{3}{2}} + c) = 0$, reducing to $(y+c)^2 - \frac{4}{9} ax^3 = 0$, is the general solution of the given equation.

EXAMPLE II. — Solve $y \dfrac{dy^2}{dx^2} + 2 x \dfrac{dy}{dx} - y = 0$.

Solving for $\dfrac{dy}{dx}$, $\dfrac{dy}{dx} = -\dfrac{x}{y} \pm \dfrac{\sqrt{x^2+y^2}}{y}$ whence $\dfrac{x\,dx + y\,dy}{\pm \sqrt{x^2+y^2}} = dx$.

Integrating, $\pm (x^2 + y^2)^{\frac{1}{2}} = x + c$. Hence the general solution is $\{(x+c) + (x^2+y^2)^{\frac{1}{2}}\} \{(x+c) - (x^2+y^2)^{\frac{1}{2}}\} = 0$, reducing to $y^2 = 2cx + c^2$.

CASE II. — Suppose the equation (1) $f(x, y, p) = 0$, where p stands for $\dfrac{dy}{dx}$, can be put into the form (2) $y = F(x, p)$. Forming the x-derivative of (2) gives an equation of the form

(3) $p = F_1\left(x, p, \dfrac{dp}{dx}\right)$. If the primitive of (3) is (4) $f_1(x, p, c) = 0$, the elimination of p from (1) and (4) is the solution of (1).

In like manner if (1) $f(x, y, p) = 0$ can be put into the form (2) $x = F(y, p)$, the y-derivative of (2) is (3) $\dfrac{1}{p} = F_1\left(y, p, \dfrac{dp}{dy}\right)$. If the primitive of (3) is (4) $f_1(y, p, c) = 0$, the elimination of p from (1) and (4) is the primitive of (1).

EXAMPLE. — Solve $y = p^2 + 2p^3$.

The x-derivative of this equation is $p = 2p\dfrac{dp}{dx} + 6p^2\dfrac{dp}{dx}$, whence $c + x = 2p + 3p^2$. Hence $x = 2p + 3p^2 - c$ and $y = p^2 + 2p^3$ for every value of p determine a pair of corresponding values of function and variable of the solution of the given equation.

CASE III. — If the equation $F(x, p) = 0$ cannot be readily solved for x or p, try the substitution $p = xz$.

EXAMPLE. — Solve $x^3 + \dfrac{dy^3}{dx^3} - ax\dfrac{dy}{dx} = 0$.

The substitution $p = xz$ gives $x^3 + x^3z^3 - ax^2z = 0$, whence

$$x = \dfrac{az}{1+z^3}. \quad \text{Now } p = \dfrac{dy}{dx} = \dfrac{dy}{dz}\dfrac{dz}{dx} = xz,$$

whence $\dfrac{dy}{dz} = \dfrac{az^2}{1+z^3}\dfrac{d}{dz}\left(\dfrac{az}{1+z^3}\right) = \dfrac{a^2(z^2 - 2z^5)}{(1+z^3)^3}$,

and by integration $y = \tfrac{1}{6}a^2\dfrac{2z^2 - 1}{(1+z^3)^2} + \tfrac{1}{3}a^2\dfrac{1}{1+z^3} + C$.

Function y and variable x are now expressed in terms of the same quantity z.

CASE IV. — If the equation $F(x, y, p) = 0$ is homogeneous in x and y, substitute $y = xz$.

EXAMPLE. — Solve $(2p+1)x^{\frac{1}{2}}y = x^{\frac{3}{2}}p^2 + 2y^{\frac{3}{2}}$.

The substitution $y = zx$, $\dfrac{dy}{dx} = z + x\dfrac{dz}{dx}$ gives $\dfrac{dx}{x} \pm \dfrac{dz}{z - z^{\frac{1}{2}}} = 0$,

whence $cx + (z^{\frac{1}{2}} - 1)^2 = 0$, and $cx + (z^{\frac{1}{2}} - 1)^{-2} = 0$ or

$$cx^2 + (y^{\frac{1}{2}} - x^{\frac{1}{2}})^2 = 0 \text{ and } c + (y^{\frac{1}{2}} - x^{\frac{1}{2}})^{-2} = 0.$$

CASE V. — Clairault's equation, $y = px + f(p)$.

The x-derivative of this equation is $p = p + x\dfrac{dp}{dx} + f'(p)\dfrac{dp}{dx}$, which reduces to (1) $\dfrac{dp}{dx}\{x + f'(p)\} = 0$. Equation (1) is satisfied by (2) $\dfrac{dp}{dx} = 0$ or (3) $x + f'(p) = 0$. From (2) $p = c$ and the general primitive of Clairault's equation is $y = cx + f(c)$. The elimination of p from the given equation and (3) gives a singular solution of Clairault's equation.

EXAMPLE I. — Solve $y = px + p - p^3$.

The general primitive, found by substituting $p = c$, is $y = cx + c - c^3$.

EXAMPLE II. — Solve $x^2(y - px) = yp^2$.

Multiply the given equation by y and substitute $u = y^2$, $\dfrac{du}{dx} = 2y\dfrac{dy}{dx}$. There results $ux^2 - \frac{1}{2}x^3\dfrac{du}{dx} = \frac{1}{4}\dfrac{du^2}{dx^2}$. Substituting $x^2 = v$, whence $\dfrac{du}{dx} = 2x\dfrac{du}{dv}$, $u = v\dfrac{du}{dv} + \dfrac{du^2}{dv^2}$, a Clairault's equation whose primitive is $u = cv + c^2$. Hence the primitive of the given equation is $y^2 = cx^2 + c^2$.

PROBLEMS

Solve,

1. $\dfrac{dy^2}{dx^2} - 7\dfrac{dy}{dx} + 12 = 0.$

2. $4y = x^2 + p^2.$

3. $y^2 = x^2(1 + p^2).$

4. $y^2 + xyp - x^2p^2 = 0.$

5. $y = xp + \sin^{-1} p$.
7. $y^2(1 - p^2) = b$.

6. $x + \dfrac{p}{\sqrt{1+p^2}} = a$.
8. $y = px + \dfrac{m}{p}$.

9. $3p^2y^2 - 2xyp + 4y^2 - x^2 = 0$.

10. $\dfrac{dy}{dx} + 2xy = x^2 + y^2$.
14. $x^2p^2 = 1 + p^2$.

11. $xy^2(p^2 + 2) = 2py^3 + x^3$.
15. $x\dfrac{dy^2}{dx^2} = 1 - x$.

12. $x^2 + y = p^2$.
16. $x^2p^2 + 3xyp + 2y^2 = 0$.

13. $\dfrac{dy^2}{dx^2} - \dfrac{a}{x} = 0$.
17. $\dfrac{dy^2}{dx^2}(x^2 + 1)^3 = 1$.

18. $y\dfrac{dy^2}{dx^2} + 2x\dfrac{dy}{dx} - y = 0$.

19. $(x^2 - 1)\dfrac{dy^2}{dx^2} - 2xy\dfrac{dy}{dx} = 1 - y^2$. Show that the singular solution is $x^2 + y^2 = 1$.

20. Find a curve such that the area bounded by the tangent and the coordinate axes is always a^2.

Art. 77. — Ordinary Equations in Three Variables

If the differential equation (1) $P\,dx + Q\,dy + R\,dz = 0$ can be solved, it may be rendered exact by some factor μ. If (2) $u = f(x, y, z) = 0$ is the solution of (1), $\dfrac{\partial u}{\partial x}dx + \dfrac{\partial u}{\partial y}dy + \dfrac{\partial u}{\partial z}dz = 0$ and $\mu P\,dx + \mu Q\,dy + \mu R\,dz = 0$ are identical,

whence $\quad \dfrac{\partial u}{\partial x} \equiv \mu P, \quad \dfrac{\partial u}{\partial y} \equiv \mu Q, \quad \dfrac{\partial u}{\partial z} \equiv \mu R$.

The identities $\dfrac{\partial^2 u}{\partial x\,\partial y} \equiv \dfrac{\partial^2 u}{\partial y\,\partial x}, \quad \dfrac{\partial^2 u}{\partial x\,\partial z} \equiv \dfrac{\partial^2 u}{\partial z\,\partial x}, \quad \dfrac{\partial^2 u}{\partial y\,\partial z} \equiv \dfrac{\partial^2 u}{\partial z\,\partial y}$ lead to the identities

$$(3) \quad \mu\left(\frac{\partial P}{\partial y} - \frac{\partial Q}{\partial x}\right) \equiv Q\frac{\partial \mu}{\partial x} - P\frac{\partial \mu}{\partial y},$$

$$(4) \quad \mu\left(\frac{\partial Q}{\partial z} - \frac{\partial R}{\partial y}\right) \equiv R\frac{\partial \mu}{\partial y} - Q\frac{\partial \mu}{\partial z},$$

$$(5) \quad \mu\left(\frac{\partial R}{\partial x} - \frac{\partial P}{\partial z}\right) \equiv P\frac{\partial \mu}{\partial z} - R\frac{\partial \mu}{\partial x}.$$

Multiply (3) by R, (4) by P, (5) by Q, and add the products. There results

$$(6) \quad P\left(\frac{\partial Q}{\partial z} - \frac{\partial R}{\partial y}\right) + Q\left(\frac{\partial R}{\partial x} - \frac{\partial P}{\partial z}\right) + R\left(\frac{\partial P}{\partial y} - \frac{\partial Q}{\partial x}\right) \equiv 0,$$

the condition under which (1) can be solved.

EXAMPLE I. — Solve $(y + z)\, dx + dy + dz = 0$.

Here $P = y + z$, $Q = 1$, $R = 1$,

$$\frac{\partial P}{\partial y} = 1, \; \frac{\partial P}{\partial z} = 1, \; \frac{\partial Q}{\partial x} = 0, \; \frac{\partial Q}{\partial z} = 0, \; \frac{\partial R}{\partial x} = 0, \; \frac{\partial R}{\partial y} = 0,$$

and condition (6) is satisfied.

Considering x constant, the given equation becomes $dy + dz = 0$, whence $y + z + X = 0$, where X must be so determined that the x-derivative of $y + z + X$ is $y + z$. Hence $\frac{dX}{dx} = y + z = -X$, $\frac{1}{X}\frac{dX}{dx} = -1$, $\log X = c - x$, $X = e^{c-x}$. Finally, $y + z + e^{c-x} = 0$, the solution required.

EXAMPLE II. — Solve

$$2(y + z)\, dx + (x + 3y + 2z)\, dy + (x + y)\, dz = 0.$$

Here $P = 2(y + z)$, $Q = x + 3y + 2z$, $R = x + y$,

$$\frac{\partial P}{\partial y} = 2, \; \frac{\partial P}{\partial z} = 2, \; \frac{\partial Q}{\partial x} = 1, \; \frac{\partial Q}{\partial z} = 1, \; \frac{\partial R}{\partial x} = 1, \; \frac{\partial R}{\partial y} = 1,$$

and condition (6) is satisfied.

Considering y constant, the given equation becomes $2(y+z)\,dx + (x+y)\,dz = 0$, or $\dfrac{2\,dx}{x+y} + \dfrac{dz}{y+z} = 0$. Integrating, $\log(x+y)^2 + \log(y+z) = \log Y$, whence $(x+y)^2(y+z) = Y$, where Y must be so determined that the y-derivative of $(x+y)^2(y+z) - Y$ is $(x+3y+2z)(x+y)$. Hence $2(x+y)(y+z) + (x+y)^2 - \dfrac{dY}{dy} = (x+3y+2z)(x+y)$, $\dfrac{dY}{dy} = 0$, $Y = C$. Finally, the required solution is $(x+y)^2(y+z) = C$.

PROBLEMS

Solve,
1. $(y+a)^2\,dx + z\,dy - (y+a)\,dz = 0$.
2. $dx + dy + (x+y+z+1)\,dz = 0$.
3. $yz\,dx + zx\,dy + xy\,dz = 0$.
4. $(y+z)\,dx + (z+x)\,dy + (x+y)\,dz = 0$.
5. $zy\,dx - zx\,dy + y^2\,dz = 0$.
6. $(y^2 + yz)\,dx + (xz + z^2)\,dy + (y^2 - xy)\,dz = 0$.

CHAPTER XIII

ORDINARY DIFFERENTIAL EQUATIONS OF HIGHER ORDER

Art. 78. — Equations of Higher Order and First Degree

STANDARD I. — The primitive of an equation of the form $\frac{d^n y}{dx^n} = f(x)$ is found by n successive integrations; the primitive of an equation of the form $\frac{d^2 y}{dx^2} = f(y)$ is found by multiplying both sides by $\frac{dy}{dx}$ and integrating, then solving for $\frac{dx}{dy}$ and integrating again.

EXAMPLE. — Solve $\frac{d^2 y}{dx^2} = y$.

Multiplying by $\quad \frac{dy}{dx}, \quad \frac{dy}{dx} \cdot \frac{d\frac{dy}{dx}}{dx} = y \frac{dy}{dx}$.

Integrating, $\quad \frac{1}{2}\left(\frac{dy}{dx}\right)^2 = \frac{1}{2} y^2 + C_1$, whence $\frac{dy}{dx} = \sqrt{y^2 + 2 C_1}$.

Solving or $\quad \frac{dx}{dy}, \quad \frac{dx}{dy} = \frac{1}{\sqrt{y^2 + 2 C_1}}$.

Integrating, $\quad x = \log \{y + \sqrt{y^2 + 2 C_1}\} + C_2$.

STANDARD II. — Equations of the form $\frac{d^2 y}{dx^2} + f_1(x) \frac{dy}{dx} = f_2(x)$ become linear by the substitution $\frac{dy}{dx} = p$, $\frac{d^2 y}{dx^2} = \frac{dp}{dx}$.

EXAMPLE.—Solve $(1-x^2)\dfrac{d^2y}{dx^2} - x\dfrac{dy}{dx} = 0$.

Substituting $\dfrac{dy}{dx} = p$, $\dfrac{d^2y}{dx^2} = \dfrac{dp}{dx}$, the given equation becomes $\dfrac{1}{p}\dfrac{dp}{dx} = \dfrac{x}{1-x^2}$. Integrating, $\dfrac{dy}{dx} = C_1(1-x^2)^{-\frac{1}{2}}$. Integrating again, $y = C_1 \sin^{-1} x + C_2$.

STANDARD III.—The equation

$$P_0 \frac{d^n y}{dx^n} + P_1 \frac{d^{n-1} y}{dx^{n-1}} + P_2 \frac{d^{n-2} y}{dx^{n-2}} + \cdots P_n y = Q,$$

linear in y and its derivatives, the coefficients P_0, P_1, P_2, $\cdots Q$ being independent of y, is called the linear equation of order n. Suppose the coefficients P_0, P_1, P_2, $\cdots P_n$ constant and $Q = 0$. When $n = 1$, the equation becomes $P_0 \dfrac{dy}{dx} + P_1 y = 0$ and $y = e^{-c\frac{P_1}{P_0}x}$, which has the general form $y = e^{mx}$. Substituting $y = e^{mx}$ in

(1) $\quad P_0 \dfrac{d^n y}{dx^n} + P_1 \dfrac{d^{n-1} y}{dx^{n-1}} + \cdots P_n y = 0$

there results (2) $\quad P_0 m^n + P_1 m^{n-1} + P_2 m^{n-2} + \cdots P_n = 0$, which shows that $y = e^{mx}$ is a solution of (1) for the n values of m which are roots of (2). Representing these n roots by

$m_1, m_2, m_3, m_4, \cdots m_n$, $y_1 = e^{m_1 x}$, $y_2 = e^{m_2 x}$, $y_3 = e^{m_3 x}, \cdots y_n = e^{m_n x}$

are solutions of (1). Consequently

(3) $\quad y = C_1 \cdot e^{m_1 x} + C_2 \cdot e^{m_2 x} + C_3 \cdot e^{m_3 x} + \cdots C_n \cdot e^{m_n x}$,

where $C_1, C_2, C_3, \cdots C_n$ are arbitrary constants, is a solution of (1). Since (3) contains n arbitrary constants, it is the general solution of (1). The values of these n constants become known if the values of y and its first $n-1$ derivatives are known for some value of x.

DIFFERENTIAL EQUATIONS OF HIGHER ORDER

If the roots of equation (2) are real and unequal, (3) is a satisfactory form of the solution of (1).

If equation (2) has pairs of conjugate imaginary roots, $m_1 = a + b\sqrt{-1}$, $m_2 = a - b\sqrt{-1}$, or $m_1 = a + ib$, $m_2 = a - ib$, the corresponding terms of (3) are

$$C_1 \cdot e^{(a+ib)x} + C_2 \cdot e^{(a-ib)x} = e^{ax}(C_1 e^{ibx} + C_2 e^{-ibx})$$
$$= e^{ax}\{C_1(\cos bx + i\sin bx) + C_2(\cos bx - i\sin bx)\}$$

by Problem 9, Art. 68. This result may be written

$$e^{ax}\{(C_1 + C_2)\cos bx + i(C_1 - C_2)\sin bx\},$$

which, by placing $C_1 = \tfrac{1}{2}(A - iB)$, $C_2 = \tfrac{1}{2}(A + iB)$, becomes $e^{ax}(A\cos bx + B\sin bx)$. Finally, placing $A = K\sin k$, $B = K\cos k$, the result becomes $Ke^{ax}\sin(k + bx)$, where K and k are arbitrary constants.

If equation (2) has equal roots $m_1 = m_2$, (3) contains less than n arbitrary constants and is no longer the general primitive. In this case write $m_2 = m_1 + h$ and the problem reduces itself to determining the form of the general primitive

$$y = C_1' \cdot e^{m_1 x} + C_2 \cdot e^{(m_1 + h)x} + C_3 \cdot e^{m_3 x} + \cdots C_n \cdot {}^{m_n x},$$

when h approaches zero. Now

$$C_1 \cdot e^{m_1 x} + C_2 \cdot e^{(m_1 + h)x} = e^{m_1 x}(C_1 + C_2 \cdot e^{hx})$$
$$= e^{m_1 x}\{C_1 + C_2(1 + hx + \frac{h^2 x^2}{2!} + \cdots)\}$$

by Maclaurin's series. When h approaches zero, this becomes $e^{m_1 x}\{(C_1 + C_2) + C_2 hx\} = e^{m_1 x}(A + Bx)$, such values being assigned to the arbitrary constants C_1 and C_2 that $C_1 + C_2 = A$, $C_2 h = B$ when h approaches zero.

If equation (2) has three equal roots, $m_1 = m_2 = m_3$, a like analysis shows that the corresponding part of the general primitive is $e^{m_1 x}(A + Bx + Cx^2)$.

If equation (2) has equal imaginary roots, for example $m_1 = m_2 = (a + ib)$ and $m_3 = m_4 = a - ib$, the corresponding terms of the general primitive are

$$e^{ax}\{(A + Bx)\cos bx + (C + Dx)\sin bx\}.$$

Equation (2) is called the auxiliary equation of the differential equation (1).

EXAMPLE I. — Solve $\dfrac{d^2y}{dx^2} + 6\dfrac{dy}{dx} + 13y = 0.$

Substituting $y = e^{mx}$, the auxiliary equation is found to be $m^2 + 6m + 13 = 0$, whence $m_1 = -3 + 2i$, $m_2 = -3 - 2i$ and $y = e^{-3x}\{A\cos 2x + B\sin 2x\}.$

EXAMPLE II. — Solve $\dfrac{d^3y}{dx^3} - 3\dfrac{d^2y}{dx^2} + 4y = 0.$

The roots of the auxiliary equation $m^3 - 3m^2 + 4 = 0$ are $-1, 2, 2$. Hence the general primitive is

$$y = C \cdot e^{-x} + (A + Bx)e^{2x}.$$

STANDARD IV. — Linear equations with second member not zero.

Form the successive x-derivatives of

$$P_0 \cdot \frac{d^n y}{dx^n} + P_1 \cdot \frac{d^{n-1}y}{dx^{n-1}} + \cdots P_n \cdot y = X$$

until either the second derivative becomes zero or the elimination of X from the given equation and the derivative becomes possible.

EXAMPLE. — Solve $\dfrac{d^2y}{dx^2} + a^2 y = \sin bx.$

The second x-derivative of this equation is

$$\frac{d^4y}{dx^4} + a^2 \cdot \frac{d^2y}{dx^2} = -b^2 \cdot \sin bx.$$

Eliminating $\sin bx$ from this derivative and the given equation there results $\dfrac{d^4y}{dx^4}+(a^2+b^2)\dfrac{d^2y}{dx^2}+a^2b^2y=0$. The auxiliary equation of this linear differential equation is

$$m^4+(a^2+b^2)m^2+a^2b^2=0,$$

whence $m_1=ai$, $m_2=-ai$, $m_3=bi$, $m_4=-bi$. The general primitive is $y=C_1\cos(ax)+C_2\sin(ax)+C_3\cos(bx)+C_4\sin(bx)$. This value of y is a solution of the given second order differential equation if $C_3a^2b-C_4a^2b^2=0$, $-C_3b^2-C_4b=1$, whence $C_3=-\dfrac{1}{1+b^2}$, $C_4=-\dfrac{1}{b(1+b^2)}$. The required solution is, therefore,

$$y=C_1\cos(ax)+C_2\sin(ax)-\dfrac{1}{1+b^2}\cos(bx)-\dfrac{1}{b(1+b^2)}\sin(bx).$$

STANDARD V. — *Change of the independent variable.*

In the equation $\dfrac{d^2y}{dx^2}+P\dfrac{dy}{dx}+Qy=0$, where P and Q are functions of x, change the independent variable from x to z by the relations $\dfrac{dy}{dx}=\dfrac{dy}{dz}\cdot\dfrac{dz}{dx}$, $\dfrac{d^2y}{dx^2}=\dfrac{d^2y}{dz^2}\left(\dfrac{dz}{dx}\right)^2+\dfrac{dy}{dz}\cdot\dfrac{d^2z}{dx^2}$. There results, (1) $\dfrac{d^2y}{dz^2}\left(\dfrac{dz}{dx}\right)^2+\dfrac{dy}{dz}\left(\dfrac{d^2z}{dx^2}+P\dfrac{dz}{dx}\right)+Qy=0$. Now determine z so that $\dfrac{d^2z}{dx^2}+P\dfrac{dz}{dx}=0$, whence (2) $z=\displaystyle\int e^{-\int P\,dx}\cdot dx$. The elimination of x from (1) and (2) gives an equation whose solution leads to the solution of the given equation.

EXAMPLE. — Solve $\dfrac{d^2y}{dx^2}+\dfrac{2x}{1+x^2}\dfrac{dy}{dx}+\dfrac{y}{(1+x^2)^2}=0$.

Here $z=\displaystyle\int e^{-\int P\,dx}\cdot dx=\int e^{-\int\frac{2x\,dx}{1+x^2}}\cdot dx=\tan^{-1}x$, whence $x=\tan z$. The transformed equation is $\dfrac{d^2y}{dz^2}+y=0$, whence $y=C_1\cdot\cos z+C_2\cdot\sin z$. Substituting for z,

$$y=C_1\cdot\cos z+C_2\cdot\sin z=C_1\dfrac{1}{\sqrt{1+x^2}}+C_2\dfrac{x}{\sqrt{1+x^2}}.$$

PROBLEMS

Solve,

1. $\dfrac{d^3y}{dx^3} = \sin^3 x.$

2. $\dfrac{d^2y}{dx^2} + 5\dfrac{dy}{dx} + 4y = 0.$

3. $x^2 \dfrac{d^4y}{dx^4} = 1.$

4. $\dfrac{d^2y}{dx^2} + \dfrac{a^2}{y^2} = 0.$

5. $\dfrac{d^2y}{dx^2} + 6\dfrac{dy}{dx} + 9y = 0.$

6. $\dfrac{d^2y}{dx^2} - m^2 y = 0.$

7. $\dfrac{d^2y}{dx^2} = x^2 \cdot \sin x.$

8. $\dfrac{d^3y}{dx^3} = \dfrac{d^2y}{dx^2} + 6\dfrac{dy}{dx}.$

9. $\dfrac{d^4y}{dx^4} + 2\dfrac{d^2y}{dx^2} - 8y = 0.$

10. $\dfrac{d^3y}{dx^3} - 3\dfrac{d^2y}{dx^2} + 4y = 0.$

11. $\dfrac{d^3y}{dx^3} = x \cdot e^x.$

12. $6\dfrac{d^2y}{dx^2} = \dfrac{dy}{dx} + y.$

13. $\dfrac{d^3y}{dx^3} - 3\dfrac{dy}{dx} + y = 0.$

14. $\dfrac{d^2y}{dx^2} - y = x + 1.$

15. $\dfrac{d^2y}{dx^2} + y = \cos x.$

16. $\dfrac{d^2y}{dx^2} - 3\dfrac{dy}{dx} + 2y = x \cdot e^x.$

17. $\dfrac{d^2y}{dx^2} + 3y = \sin(mx).$

18. $\dfrac{d^2y}{dx^2} + 4y = \cos(nx).$

19. $(1-x^2)\dfrac{d^2y}{dx^2} - x\dfrac{dy}{dx} = y.$

20. $x\dfrac{d^2y}{dx^2} - \dfrac{dy}{dx} + 4x^3 y = x^5.$

21. $\dfrac{d^2y}{dx^2} = \left(\dfrac{dy}{dx}\right)^2 + 1.$

22. $x\dfrac{d^2y}{dx^2} + \dfrac{dy}{dx} = x^n.$

Art. 79. — Symbolic Integration

If u, v, w are any symbols whatever obeying the fundamental laws of ordinary algebra, namely:

the associative law, $u + v + w \equiv (u+v)+w$, $uvw \equiv (uv)w$;
the commutative law, $u + v + w \equiv v + u + w$, $uvw \equiv vuw$;
the distributive law, $(u + v)w \equiv uw + vw$;
the law of indices, $u^m u^n \equiv u^{m+n}$;

any algebraic transformation of expressions involving u, v, w gives a valid result.

Denoting the operation of forming the first derivative by D, that of forming the second derivative by D^2, that of forming the third derivative by D^3, ..., so that $Dy \equiv \dfrac{dy}{dx}$, $D^2 y \equiv \dfrac{d^2 y}{dx^2}$, $D^3 y \equiv \dfrac{d^3 y}{dx^3}$, ...; and defining the symbol D^{-1} by the equation $D(D^{-1} y) \equiv y$, whence D^{-1} is equivalent to the symbol of integration $\int$, it follows that:

I. $D \cdot Dy \equiv D^2 y$, $\quad D \cdot D^2 y \equiv D^3 y$, $\quad D^{-1} \cdot D^2 y \equiv Dy$, $D^{-2} \cdot Dy \equiv D^{-1} y$; that is, the law of indices of algebra holds for D affected by integral exponents.

II. $(a + bD - cD^2) y \equiv \{(a + bD) - cD^2\} y$; that is, the associative law of algebra holds for D combined with constants.

III. $(aD + bD^2) y \equiv D(a + bD) y$; that is, the distributive law of algebra holds for D combined with constants.

IV. $(a + bD) y \equiv (bD + a) y$, $aDy \equiv Day$; that is, the commutative law of algebra holds for D combined with constants.

Since the operator D obeys the fundamental laws of algebra, the result of any algebraic transformation of expressions containing D and constants is valid. For example, by Maclaurin's series,
$$f(D) \equiv A_0 + A_1 \cdot D + A_2 \cdot D^2 + \cdots A_n \cdot D^n + \cdots \equiv \Sigma A_n \cdot D^n.$$
Hence $f(D) y \equiv \Sigma A_n \cdot D^n y$. If $f(D) y = X$, $y = \dfrac{1}{f(D)} X$. This is the definition of the operator $\dfrac{1}{f(D)}$ inverse to $f(D)$.

For instance,
$$(D^2 - 3D + 1)(2x^3 - 5x^2 + 7x) = 2x^3 - 23x^2 + 49x - 31.$$
Hence $\dfrac{1}{D^2 - 3D + 1}(2x^3 - 23x^2 + 49x - 31) = 2x^3 - 5x^2 + 7x.$

EXAMPLE I. — Show that $(D^2 - D - 2)y \equiv (D+1)(D-2)y.$
$(D^2 - D - 2)y \equiv \dfrac{d^2y}{dx^2} - \dfrac{dy}{dx} - 2y;\ (D-2)y \equiv \dfrac{dy}{dx} - 2y$ and
$(D+1)(D-2)y \equiv (D+1)\left(\dfrac{dy}{dx} - 2y\right) \equiv \dfrac{d^2y}{dx^2} - 2\dfrac{dy}{dx} + \dfrac{dy}{dx} - 2y$
$\equiv \dfrac{d^2y}{dx^2} - \dfrac{dy}{dx} - 2y.$

EXAMPLE II. — Show that $f(D)e^{ax} \equiv f(a)e^{ax}.$
$$f(D) \equiv \Sigma A_n D^n \text{ and } A_n D^n e^{ax} = A_n a^n e^{ax}.$$
Hence $f(D)e^{ax} = f(a)e^{ax}.$ Inversely $\dfrac{1}{f(D)}e^{ax} = \dfrac{1}{f(a)}e^{ax}.$

EXAMPLE III. — Show that $\dfrac{1}{D^2 - D - 2}y \equiv \left(\dfrac{\frac{1}{3}}{D+1} - \dfrac{\frac{1}{3}}{D-2}\right)y.$
Performing the operation $D^2 - D - 2$ on both sides of the assumed identity,
$$(D^2 - D - 2)\dfrac{1}{D^2 - D - 2}y \equiv (D^2 - D - 2)\left(\dfrac{\frac{1}{3}}{D+1} - \dfrac{\frac{1}{3}}{D-2}\right)y$$
or $y \equiv \{-\frac{1}{3}D + \frac{2}{3} + \frac{1}{3}D + \frac{1}{3}\}y \equiv y,$ which proves the assumed identity correct.

EXAMPLE IV. — $(D^2 - 2D + 2)y = 6x - 9x^2 + 2x^3.$ Find $y.$
$y = \dfrac{1}{D^2 - 3D + 2}(6x - 9x^2 + 2x^3)$
$\equiv \dfrac{1}{1 - D}(6x - 9x^2 + 2x^3) - \dfrac{1}{2 - D}(6x - 9x^2 + 2x^3)$
$\equiv (1 + D + D^2 + D^3 + D^4 + \cdots)(6x - 9x^2 + 2x^3)$
$- \left(\dfrac{1}{2} + \dfrac{D}{4} + \dfrac{D^2}{8} + \dfrac{D^3}{16} + \dfrac{D^4}{32} + \cdots\right)(6x - 9x^2 + 2x^3)$
$= x^3.$

DIFFERENTIAL EQUATIONS OF HIGHER ORDER 233

This is a particular value of y since it does not contain arbitrary constants.

EXAMPLE V. — $y = \dfrac{1}{D-2}[0]$. Find y.

This is equivalent to $(D-2)y = 0$, which is the same as the linear equation $\dfrac{dy}{dx} - 2y = 0$. Whence $y = e^{2x}$.

PROBLEMS

1. Show that $f(D)[e^{ax}X] \equiv e^{ax} \cdot f(D+a)[X]$, and consequently $\dfrac{1}{f(D)}[e^{ax}X] \equiv e^{ax}\dfrac{1}{f(D+a)}[X]$.

2. $(D^2 + D + 1)y = e^x x^3$. Show that $y = e^x\left(\dfrac{x^3}{3} - x^2 + \tfrac{4}{3}x + \dfrac{2}{9}\right)$.

3. Show that $f(D^2)\sin(mx) \equiv f(-m^2)\sin(mx)$ and consequently $\dfrac{1}{f(D^2)}\sin(mx) \equiv \dfrac{1}{f(-m^2)}\sin(mx)$.

4. Show that $f(D^2)\cos(mx) \equiv f(-m^2)\cos(mx)$ and consequently $\dfrac{1}{f(D^2)}\cos(mx) \equiv \dfrac{1}{f(-m^2)}\cos(mx)$.

5. Show that $\dfrac{1}{D(D-1)}x = -\dfrac{x^2}{2} - x$.

6. Show that $\dfrac{1}{D^2 - 4D + 4}e^{2x}\sin x = -e^{2x}\sin x$.

7. Show that $\dfrac{1}{D^3 + D^2 + D + 1}e^{2x} = \dfrac{e^{2x}}{15}$.

8. Show that $\dfrac{D}{D-1}e^x \sin x = e^x(\sin x - \cos x)$.

9. $(D-m)y = 0$. Show that $y = \dfrac{1}{D-m}[0] = e^{mx}$.

10. Show that $y = \dfrac{C}{(D-m)^2}[0] = e^{mx}(C_1 + C_2 x)$.

Art. 80.—Symbolic Solution of Linear Equations*

Writing the linear equation $P_0 \dfrac{d^n y}{dx^n} + P_1 \dfrac{d^{n-1} y}{dx^{n-1}} + \cdots P_n y = 0$ with constant coefficients and second member zero in the form

$$(P_0 D^n + P_1 D^{n-1} + P_2 D^{n-2} + \cdots P_n) y = 0,$$

$$y = \dfrac{1}{P_0 D^n + P_1 D^{n-1} + P_2 D^{n-2} + \cdots P_n}[0].$$

Factoring, $P_0 D^n + P_1 D^{n-1} + P_2 D^{n-2} + \cdots P_n$
$$\equiv (D - m_1)(D - m_2)(D - m_3) \cdots (D - m_n).$$

Decomposing into partial fractions,

$$y = \dfrac{C_1}{D - m_1}[0] + \dfrac{C_2}{D - m_2}[0] + \dfrac{C_3}{D - m_3}[0] + \cdots + \dfrac{C_n}{D - m_n}[0],$$

hence, $\quad y = C_1 e^{m_1 x} + C_2 e^{m_2 x} + C_3 e^{m_3 x} + \cdots + C_n e^{m_n x}.$

If $m_1 = m_2$, a partial fraction of the form $\dfrac{C}{(D - m_1)^2}$ occurs. The value of $\dfrac{C}{(D - m_1)^2}[0]$ is $(C_1 + C_2 x) e^{m_1 x}$.

In general, if $m_1 = m_2 = m_3 = \cdots m_r$, a fraction of the form $\dfrac{C}{(D - m_1)^r}$ occurs. The value of $\dfrac{C}{(D - m_1)^r}[0]$ is

$$(C_1 + C_2 x + C_3 x^2 + \cdots C_{r-2} x^{r-3} + C_{r-1} x^{r-2} + C_r x^{r-1}) e^{m_1 x}.$$

If $m_1 = m_2 = a + bi$, $m_3 = m_4 = a - bi$, the corresponding terms of the solution of the differential equation are

$$(C_1 + C_2 x) e^{a + bi} + (C_3 + C_4) e^{a - bi},$$

reducing to $(A_1 x + B_1) \cos(bx) + (A_2 + B_2 x) \sin(bx).$

The solution of (1) $P_0 \dfrac{d^n y}{dx^n} + P_1 \dfrac{d^{n-1} y}{dx^{n-1}} + \cdots + P_n y = X$, with

---

*Maclaurin introduced the symbolic solution of differential equations.

coefficients constant and right-hand member a function of x, is $y = Y + u$, where Y, called the complementary solution of (1), is the solution of $P_0 \dfrac{d^n y}{dx^n} + P_1 \dfrac{d^{n-1} y}{dx^{n-1}} + \cdots + P_n y = 0$, and u is any particular solution of (1). For substituting $y = Y + u$ in (1),
$$\left(P_0 \frac{d^n Y}{dx^n} + P_1 \frac{d^{n-1} Y}{dx^{n-1}} + \cdots + P_n Y \right)$$
$$+ \left(P_0 \frac{d^n u}{dx^n} + P_1 \frac{d^{n-1} u}{dx^{n-1}} + \cdots P_n u \right) = X,$$
which is a true equation by the hypothesis.

EXAMPLE I. — Solve $\dfrac{d^4 y}{dx^4} - y = x^4$.

Writing the complementary equation $(D^4 - 1) Y = 0$, or $(D+1)(D-1)(D+\sqrt{-1})(D-\sqrt{-1}) Y = 0$, the complementary solution is found to be $Y = C_1 e^x + C_2 e^{-x} + C_3 \sin x + C_4 \cos x$. The particular solution is
$$u = \frac{1}{D^4 - 1} x^4 \equiv (-1 - D^4 - D^8 - \cdots) x^4 = -x^4 - 24.$$

Hence the general solution of the given equation is
$$y = C_1 e^x + C_2 e^{-x} + C_3 \sin x + C_4 \cos x - x^4 - 24.$$

EXAMPLE II. — Solve $\dfrac{d^2 y}{dx^2} - 2 \dfrac{dy}{dx} + y = x^2 e^{3x}$.

The solution of the complementary equation
$$(D^2 - 2D + 1) Y = 0 \text{ is } Y = e^x (C_1 + C_2 x).$$

The particular solution is
$$u = \frac{1}{D^2 - 2D + 1} (x^2 e^{3x})$$
$$\equiv e^{3x} \frac{1}{(D+3)^2 - 2(D+3) + 1} x^2 \equiv e^{3x} \frac{1}{(2+D)^2} x^2$$
$$\equiv e^{3x} \left(\frac{1}{4} - \frac{D}{4} + \frac{3 D^2}{16} + \cdots \right) x^2 = \tfrac{1}{8} e^{3x} (2 x^2 - 4 x + 3).$$

The general solution is
$$y = e^x(C_1 + C_2 x) + \tfrac{1}{8} e^{3x}(2x^2 - 4x + 3).$$

EXAMPLE III. — Solve $\dfrac{d^2y}{dx^2} + \dfrac{dy}{dx} + y = \sin(2x)$.

The solution of the complementary equation
$$(D^2 + D + 1)Y = 0 \text{ is } Y = e^{-\frac{x}{2}}\left(C_1 \cos\frac{\sqrt{3}}{2}x + C_2 \sin\frac{\sqrt{3}}{2}x\right).$$

The particular solution is
$$u = \frac{1}{D^2 + D + 1}\sin(2x) \equiv \frac{1}{D-3}\sin 2x \equiv \frac{D+3}{D^2-9}\sin 2x$$
$$\equiv -\tfrac{1}{13}(D+3)\sin 2x = -\tfrac{1}{13}(2\cos 2x + 3\sin 2x).$$

Hence the general solution is
$$y = e^{-\frac{x}{2}}\left(C_1 \cos\frac{\sqrt{3}}{2}x + C_2 \sin\frac{\sqrt{3}}{2}x\right) - \tfrac{1}{13}(2\cos 2x + 3\sin 2x).$$

PROBLEMS

Solve,

1. $\dfrac{d^2y}{dx^2} + y = 1 + x + x^2$.

2. $\dfrac{d^2y}{dx^2} - 2\dfrac{dy}{dx} + y = e^x$.

3. $\dfrac{d^2y}{dx^2} + y = \cos x$.

4. $\dfrac{d^2y}{dx^2} - 3\dfrac{dy}{dx} + 2y = xe^{2x}$.

5. $\dfrac{d^2y}{dx^2} + 4y = x \sin^2 x$.

6. $\dfrac{d^3y}{dx^3} - 2\dfrac{dy}{dx} + 4y = e^x \cos x$.

7. $\dfrac{d^2y}{dx^2} + 3y = \sin\dfrac{2\pi}{T}x$.

8. $\dfrac{d^2y}{dx^2} + 4y = \cos(nx)$.

9. $\dfrac{d^2y}{dx^2} + y = x \sin 2x$.

10. $\dfrac{d^2y}{dx^2} + 4y = 2x^3 \sin^2 x$.

11. $\dfrac{d^2y}{dx^2} + 2\dfrac{dy}{dx} + 2y = e^x \sin x + \cos x$.

12. $\dfrac{d^4y}{dx^4} + 2\dfrac{d^2y}{dx^2} + y = x^2 \cos x$.

Art. 81. — Systems of Simultaneous Differential Equations

Let $P\dfrac{dy}{dx} = Q$, $P_1\dfrac{dz}{dx} = Q_1$, where x is the independent variable and y and z are the dependent variables, and P, Q, P_1, Q_1 are functions of x, y, z. It may be possible by combining the given equations with their derivatives to obtain an equation in which one dependent variable and its derivatives do not appear.

EXAMPLE. — Solve $\dfrac{dx}{dt} - 7x + y = 0$, $\dfrac{dy}{dt} - 2x - 5y = 0$.

Form the t-derivative of the first equation,

$$\frac{d^2x}{dt^2} - 7\frac{dx}{dt} + \frac{dy}{dt} = 0$$

and eliminate y and $\dfrac{dy}{dt}$ from this equation and the two given equations. There results the linear equation

$$\frac{d^2x}{dt^2} - 12\frac{dx}{dt} + 37x = 0,$$

whence $x = e^{6t}(C_1 \cdot \cos t + C_2 \cdot \sin t)$. Substituting this value of x in the first of the given equations,

$$y = e^{6t}\{(C_1 - C_2)\cos t + (C_1 + C_2)\sin t\}.$$

If the given equations can be written in the form

$$\frac{dx}{P} = \frac{dy}{Q} = \frac{dz}{R},$$

each of these equal ratios equals $\dfrac{l\,dx + m\,dy + n\,dz}{lP + mQ + nR}$, and the last ratio implies the same relation between x, y, and z as the three given ratios. If l, m, n can be so determined that

$lP + mQ + nR \equiv 0$, it follows that $l\,dx + m\,dy + n\,dz = 0$. The integral of the last equation is an integral of the given system of equations.

EXAMPLE. — Solve $\dfrac{x\,dx}{y^2z} = \dfrac{dy}{xz} = \dfrac{dz}{y^2}$.

Write $\dfrac{x\,dx}{y^2z} = \dfrac{dy}{xz} = \dfrac{dz}{y^2} = \dfrac{lx\,dx + m\,dy + n\,dz}{ly^2z + mxz + ny^2}$.

Placing $l = -x$, $m = y^2$, $n = 0$, $ly^2z + mxz + ny^2 \equiv 0$, whence $-x^2\,dx + y^2\,dy = 0$ and $y^3 - x^3 = C_1$.

Placing $l = -1$, $m = 0$, $n = z$, $ly^2z + mxz + ny^2 \equiv 0$, whence $-x\,dx + z\,dz = 0$ and $z^2 - x^2 = C_2$.

PROBLEMS

Solve, 1. $\dfrac{dx}{dt} + \dfrac{dy}{dt} + 2x + y = 0$, $\dfrac{dx}{dt} + 5x + 3y = 0$.

2. $\dfrac{dx}{dt} + 5x - 2y = e^t$, $\dfrac{dy}{dt} - x + 6y = e^{2t}$.

3. $\dfrac{dx}{x^2} = \dfrac{dy}{y^2} = \dfrac{dz}{xy}$.

4. $-dx = \dfrac{dy}{3y + 4z} = \dfrac{dz}{2y + 5z}$.

5. $\dfrac{d^2x}{dt^2} - 3x + 4y + 3 = 0$, $\dfrac{d^2y}{dt^2} + x + y + 5 = 0$.

6. $\dfrac{d^2x}{dt^2} + 3\dfrac{dy}{dt} + 16x = 0$, $\dfrac{d^2y}{dt^2} - 5\dfrac{dx}{dt} + 9y = 0$.

CHAPTER XIV

PARTIAL DIFFERENTIAL EQUATIONS

Art. 82. — Formation of Partial Differential Equations

EXAMPLE I. — Form the partial differential equation of the system of spheres (1) $(x-a)^2 + (y-b)^2 + z^2 = R^2$, whose centers $(a, b, 0)$ lie in the XY-plane and whose radius is constant.

Consider z the dependent, x and y the independent variables, and denote $\dfrac{\partial z}{\partial x}$ by p, $\dfrac{\partial z}{\partial y}$ by q.

Differentiating (1) partially with respect to x,

$$(2) \quad (x-a) + zp = 0;$$

differentiating (1) partially with respect to y,

$$(3) \quad (y-b) + zq = 0.$$

Eliminating a and b from (1), (2), (3), $z^2(1 + p^2 + q^2) = R^2$, the partial differential equation required.

EXAMPLE II. — Form the partial differential equation of the expression $z = \phi_1(y + ax) + \phi_2(y - ax)$, where ϕ_1 and ϕ_2 are arbitrary functions.

Writing the given expression $z = \phi_1(v_1) + \phi_2(v_2)$, which requires that $v_1 = y + ax$, $v_2 = y - ax$, and differentiating twice in succession partially with respect to x and also with respect to y,

$$\frac{\partial z}{\partial x} = a\phi'_1(v_1) - a\phi'_2(v_2), \quad \frac{\partial^2 z}{\partial x^2} = a^2 \cdot \phi''_1(v_1) + a^2 \cdot \phi''_2(v_2),$$

$$\frac{\partial z}{\partial y} = \phi'_1(v_1) + \phi'_2(v_2), \quad \frac{\partial^2 z}{\partial y^2} = \phi''_1(v_1) + \phi''_2(v_2).$$

By division, $\frac{\partial^2 z}{\partial x^2} = a^2 \cdot \frac{\partial^2 z}{\partial y^2}$, the partial differential equation required.

Observe that partial differential equations result either from the elimination of arbitrary constants from a function and its successive partial derivatives, or from the elimination of arbitrary functions from an expression and its successive derivatives.

PROBLEMS

Form the partial differential equations of the following expressions:

1. $z = ax + \frac{y}{a} + b.$
2. $z = ax + a^2 y^2 + b.$
3. $z = ax + by + ab.$
4. $z = \phi(y + mx).$
5. $y - bz = \phi(x - az).$
6. $z + ay + bx = \phi\{(x - \alpha)^2 + (y - \beta)^2 + z^2\}.$

ART. 83. — PARTIAL DIFFERENTIAL EQUATIONS OF FIRST ORDER

STANDARD I. — Equations of the form $F(p, q) = 0$.

Try a solution of the form $z = ax + by + c$. Since $p = a$ and $q = b$, $z = ax + by + c$ is a solution of $F(p, q) = 0$ if $F(a, b) = 0$. Denoting by $f(a)$ the value of b obtained from the equation $F(a, b) = 0$, the solution of $F(p, q) = 0$ is $z = ax + f(a) \cdot y + c$.

EXAMPLE. — Solve $p^2 + q^2 = m^2$.

Here $a^2 + b^2 = m^2$, whence $b = \sqrt{m^2 - a^2}$, and the solution is $z = ax + \sqrt{m^2 - a^2} \cdot y + c$.

STANDARD II. — Equations of the form $F(z, p, q) = 0$.

Try a solution of the form $z = \phi(x + ay)$. Writing $z = \phi(v)$, $v = x + ay$, $p = \dfrac{dz}{dv} \dfrac{\partial v}{\partial x} = \dfrac{dz}{dv}$ and $q = \dfrac{dz}{dv} \dfrac{\partial v}{\partial y} = a \dfrac{dz}{dv}$. By substituting these values of p and q the given equation becomes $F\left(z, \dfrac{dz}{dv}, a\dfrac{dz}{dv}\right) = 0$, an ordinary differential equation whose solution leads to the solution of the given partial differential equation.

EXAMPLE. — Solve $9(p^2 z + q^2) = 4$.

Writing $z = \phi(v)$, $v = x + ay$, whence $p = \dfrac{dz}{dv}$ and $q = a\dfrac{dz}{dv}$, the given equation becomes

$$9\left(z\dfrac{dz^2}{dv^2} + a^2 \dfrac{dz^2}{dv^2}\right) = 4, \text{ and } \dfrac{dz}{dv} = \dfrac{2}{3}\dfrac{1}{\sqrt{z + a^2}}.$$

Integrating, $\tfrac{2}{3}(v + c) = \tfrac{2}{3}(z + a^2)^{\frac{3}{2}}$ or $(x + ay + c)^2 = (z + a^2)^3$.

STANDARD III. — Equations of the form $F_1(x, p) = F_2(y, q)$.

Assume (1) $F_1(x, p) = a$, (2) $F_2(y, q) = a$, where a is an arbitrary constant. Integrating (1), $z = f_1(x, a) + Y$, where Y represents the terms of z which do not contain x; integrating (2), $z = f_2(y, a) + X$, where X represents the terms of z which do not contain y. Hence $z = f_1(x, a) + f_2(y, a) + C$ is the required solution.

EXAMPLE. — Solve $p^2 + q^2 = x + y$.

Write this equation $p^2 - x = y - q^2 = a$, whence

$$p = \frac{\partial z}{\partial x} = (a+x)^{\frac{1}{2}} \text{ and } q = \frac{\partial z}{\partial y} = (y-a)^{\frac{1}{2}}.$$

Integrating, $z = \frac{2}{3}(a+x)^{\frac{3}{2}} + Y$, $z = \frac{2}{3}(y-a)^{\frac{3}{2}} + X$,

and $\qquad z = \frac{2}{3}(a+x)^{\frac{3}{2}} + \frac{2}{3}(y-a)^{\frac{3}{2}} + C$,

the required solution.

STANDARD IV. — The analogue of Clairault's equation $z = px + qy + \phi(p, q)$. The solution of this equation is $z = ax + by + \phi(a, b)$, for $\frac{\partial z}{\partial x} = p = a$, $\frac{\partial z}{\partial y} = q = b$.

EXAMPLE. — Solve $z = px + qy + (1 + p^2 + q^2)^{\frac{1}{2}}$.

The solution is $z = ax + by + (1 + a^2 + b^2)^{\frac{1}{2}}$.

STANDARD V. — Lagrange's solution of (1) $P\frac{\partial z}{\partial x} + Q\frac{\partial z}{\partial y} = R$, where P, Q, R are functions of x, y, z.

Suppose (2) $u = F(x, y, z) = a$ to be a solution of (1). Differentiating (2) partially with respect to x and y,

(3) $\dfrac{\partial u}{\partial x} + \dfrac{\partial u}{\partial z}\dfrac{\partial z}{\partial x} = 0$, (4) $\dfrac{\partial u}{\partial y} + \dfrac{\partial u}{\partial z}\dfrac{\partial z}{\partial y} = 0$.

Solving (3) and (4) for $\frac{\partial z}{\partial x}$ and $\frac{\partial z}{\partial y}$ and substituting in (1), there results (5) $P\dfrac{\partial u}{\partial x} + Q\dfrac{\partial u}{\partial y} + R\dfrac{\partial u}{\partial z} = 0$. Hence a solution of (1) is also a solution of (5), and conversely.

Writing the system of equations

$$\frac{dx}{P} = \frac{dy}{Q} = \frac{dz}{R} = \frac{\frac{\partial u}{\partial x}dx + \frac{\partial u}{\partial y}dy + \frac{\partial u}{\partial z}dz}{P\frac{\partial u}{\partial x} + Q\frac{\partial u}{\partial y} + R\frac{\partial u}{\partial z}},$$

it is evident that if $u = a$ is a solution of the system of equations $\dfrac{dx}{P} = \dfrac{dy}{Q} = \dfrac{dz}{R}$, it must also be a solution of $P\dfrac{\partial u}{\partial x} + Q\dfrac{\partial u}{\partial y} + R\dfrac{\partial u}{\partial z} = 0$, and consequently of $P\dfrac{\partial z}{\partial x} + Q\dfrac{\partial z}{\partial y} = R$.

If $u = a$ and $v = b$ are two independent solutions of the system of equations $\dfrac{dx}{P} = \dfrac{dy}{Q} = \dfrac{dz}{R}$, $f(u, v) = 0$, where f represents an arbitrary function, is also a solution of (1). For

$$P\dfrac{\partial}{\partial x}f(u,v) + Q\dfrac{\partial}{\partial y}f(u,v) + R\dfrac{\partial}{\partial z}f(u,v)$$

$$\equiv P\dfrac{\partial}{\partial u}f(u,v)\cdot\dfrac{\partial u}{\partial x} + P\dfrac{\partial}{\partial v}f(u,v)\cdot\dfrac{\partial v}{\partial x} + Q\dfrac{\partial}{\partial u}f(u,v)\cdot\dfrac{\partial u}{\partial y}$$

$$+ Q\dfrac{\partial}{\partial v}f(u,v)\cdot\dfrac{\partial v}{\partial y} + R\dfrac{\partial}{\partial u}f(u,v)\cdot\dfrac{\partial u}{\partial z} + R\dfrac{\partial}{\partial v}f(u,v)\cdot\dfrac{\partial v}{\partial z}$$

$$\equiv \dfrac{\partial}{\partial u}f(u,v)\left\{P\dfrac{\partial u}{\partial x} + Q\dfrac{\partial u}{\partial y} + R\dfrac{\partial u}{\partial z}\right\}$$

$$+ \dfrac{\partial}{\partial v}f(u,v)\left\{P\dfrac{\partial v}{\partial x} + Q\dfrac{\partial v}{\partial y} + R\dfrac{\partial v}{\partial z}\right\} = 0 \text{ by hypothesis.}$$

The solution $f(u, v) = 0$ may be written $u = \phi(v)$, where ϕ represents an arbitrary function.

EXAMPLE. — Solve $xz \cdot \dfrac{\partial z}{\partial x} + yz \cdot \dfrac{\partial z}{\partial y} = xy$.

The system of equations $\dfrac{dx}{xz} = \dfrac{dy}{yz} = \dfrac{dz}{xy}$ is satisfied by $\dfrac{x}{y} = a$ and $xy - z^2 = b$. Hence the general solution of the given equation is $xy - z^2 = \phi\left(\dfrac{x}{y}\right)$.

PROBLEMS

Solve,

1. $pq = k$.   2. $q = xp + p^2$.   3. $z = px + qy + pq$.

4. $p^{\frac{1}{2}} + q^{\frac{1}{2}} = 2x$.

5. $p^2 + q^2 = npq$.

6. $p^2 = z^2(1 - pq)$.

7. $p(1 + q) = qz$.

8. $z = px + qy + 3p^{\frac{1}{3}}q^{\frac{1}{3}}$.

9. $z\dfrac{\partial z}{\partial x} + y\dfrac{\partial z}{\partial y} = x$.

10. $x\dfrac{\partial z}{\partial x} + z\dfrac{\partial z}{\partial y} + y = 0$.

11. $x^2\dfrac{\partial z}{\partial x} - xy\dfrac{\partial z}{\partial y} + y^2 = 0$.

12. $x^2\dfrac{\partial z}{\partial x} + y^2\dfrac{\partial z}{\partial y} = z^2$.

ART. 84. — LINEAR EQUATIONS OF HIGHER ORDER

STANDARD I. — All derivatives of the same order.

EXAMPLE. — Solve $\dfrac{\partial^2 z}{\partial x^2} + 3\dfrac{\partial^2 z}{\partial x\,\partial y} + 2\dfrac{\partial^2 z}{\partial y^2} = 0$.

Assume $z = \phi(y + mx)$, where ϕ represents an arbitrary function. Writing $z = \phi(v)$ and $v = y + mx$, $\dfrac{\partial z}{\partial x} = m\phi'(v)$, $\dfrac{\partial^2 z}{\partial x^2} = m^2\phi''(v)$, $\dfrac{\partial^2 z}{\partial x\,\partial y} = m\phi''(v)$, $\dfrac{\partial z}{\partial y} = \phi'(v)$, $\dfrac{\partial^2 z}{\partial y^2} = \phi''(v)$. Substituting in the given equation, $(m^2 + 3m + 2)\phi''(v) = 0$.

Hence $z = \phi(y + mx)$ is a solution of $\dfrac{\partial^2 z}{\partial x^2} + 3\dfrac{\partial^2 z}{\partial x\,\partial y} + 2\dfrac{\partial^2 z}{\partial y^2} = 0$ if $m^2 + 3m + 2 = 0$, that is if $m = -1$, $m = -2$. The required solution is therefore $z = \phi_1(y - x) + \phi_2(y - 2x)$.

STANDARD II. — General linear equation.

EXAMPLE. — Solve $\dfrac{\partial^2 z}{\partial x^2} - \dfrac{\partial^2 z}{\partial y^2} - 3\dfrac{\partial z}{\partial x} + 3\dfrac{\partial z}{\partial y} = 0$.

Assume $z = e^{mx+ny}$, whence $\dfrac{\partial z}{\partial x} = me^{mx+ny}$, $\dfrac{\partial^2 z}{\partial x^2} = m^2 e^{mx+ny}$, $\dfrac{\partial z}{\partial y} = ne^{mx+ny}$, $\dfrac{\partial^2 z}{\partial y^2} = n^2 e^{mx+ny}$.

Substituting in the given equation,

$$(m^2 - n^2 - 3m + 3n)e^{mx+ny} = 0.$$

Hence $z = e^{mx+ny}$ is a solution of the given differential equation if $m^2 - n^2 - 3m + 3n = 0$. Solving this equation for m, $m = n$, $m = 3 - n$. Hence $z_1 = e^{n(x+y)}$ and $z_2 = e^{3x} \cdot e^{n(y-x)}$ are solutions of the given equation for all values of n, and in general $z = \Sigma A_n e^{n(x+y)} + e^{3x} \Sigma B_n e^{n(y-x)}$, where n, A_n, and B_n are arbitrary constants is a solution of the given equation. Since $\Sigma A_n e^{n(x+y)}$ and $\Sigma B_n e^{n(y-x)}$ are arbitrary functions of $x + y$ and $y - x$ respectively, this solution may be written

$$z = \phi_1(x+y) + e^{3x}\phi_2(y-x).$$

PROBLEMS

Solve,

1. $\dfrac{\partial^2 z}{\partial x^2} + 5\dfrac{\partial^2 z}{\partial x \partial y} + 6\dfrac{\partial^2 z}{\partial y^2} = 0.$

3. $\dfrac{\partial^2 y}{\partial t^2} = a^2 \dfrac{\partial^2 y}{\partial x^2}.$

2. $\dfrac{\partial^2 z}{\partial x^2} - \dfrac{\partial^2 z}{\partial x \partial y} + \dfrac{\partial z}{\partial x} - \dfrac{\partial z}{\partial y} = 0.$

4. $\dfrac{\partial^2 z}{\partial x^2} - \dfrac{\partial^2 z}{\partial x \partial y} + \dfrac{\partial z}{\partial y} - z = 0.$

5. $2\dfrac{\partial^2 z}{\partial x^2} - 3\dfrac{\partial^2 z}{\partial x \partial y} - 2\dfrac{\partial^2 z}{\partial y^2} = 0.$

NEW AMERICAN EDITION OF

HALL AND KNIGHT'S ALGEBRA,

FOR COLLEGES AND SCHOOLS.

Revised and Enlarged for the Use of American Schools
and Colleges.

By FRANK L. SEVENOAK, A.M.,

*Assistant Principal of the Academic Department, Stevens
Institute of Technology.*

Half leather. 12mo. $1.10.

JAMES LEE LOVE, Instructor of Mathematics, Harvard University, Cambridge, Mass.: — Professor Sevenoak's revision of the Elementary Algebra is an excellent book. I wish I could persuade all the teachers fitting boys for the Lawrence Scientific School to use it.

VICTOR C. ALDERSON, Professor of Mathematics, Armour Institute, Chicago, Ill.: — We have used the English Edition for the past two years in our Scientific Academy. The new edition is superior to the old, and we shall certainly use it. In my opinion it is the best of all the elementary algebras.

AMERICAN EDITION OF

ALGEBRA FOR BEGINNERS.

By H. S. HALL, M.A., and S. R. KNIGHT.

REVISED BY

FRANK L. SEVENOAK, A.M.,

*Assistant Principal of the Academic Department, Stevens
Institute of Technology.*

16mo. Cloth. 60 cents.

An edition of this book containing additional chapters on Radicals and
the Binomial Theorem will be ready shortly.

JAMES S. LEWIS, Principal University School, Tacoma, Wash.: — I have examined Hall and Knight's "Algebra for Beginners" as revised by Professor Sevenoak, and consider it altogether the best book for the purpose intended that I know of.

MARY McCLUN, Principal Clay School, Fort Wayne, Indiana: — I have examined the Algebra quite carefully, and I find it the best I have ever seen. Its greatest value is found in the simple and clear language in which all its definitions are expressed, and in the fact that each new step is so carefully explained. The examples in each chapter are well selected. I wish all teachers who teach Algebra might be able to use the "Algebra for Beginners."

THE MACMILLAN COMPANY,

66 FIFTH AVENUE, NEW YORK.

AMERICAN EDITION

OF

LOCK'S
TRIGONOMETRY FOR BEGINNERS,
WITH TABLES.

Revised for the Use of Schools and Colleges

By JOHN ANTHONY MILLER, A.M.,
Professor of Mechanics and Astronomy at the Indiana University.

8vo. Cloth. $1.10 net.

IN PREPARATION.

AMERICAN EDITION

OF

HALL and KNIGHT'S
ELEMENTARY TRIGONOMETRY,
WITH TABLES.

By H. S. HALL, M.A., and S. R. KNIGHT, B.A.

Revised and Enlarged for the Use of American
Schools and Colleges

By FRANK L. SEVENOAK, A.M.,
*Assistant Principal of the Academic Department, Stevens
Institute of Technology.*

THE MACMILLAN COMPANY,
66 FIFTH AVENUE, NEW YORK.

ELEMENTARY SOLID GEOMETRY.

BY

HENRY DALLAS THOMPSON, D.Sc., Ph.D.,

Professor of Mathematics in Princeton University.

12mo. Cloth. $1.10, net.

This is an elementary work on Geometry, brief and interesting, well conceived, and well written. — *School of Mines Quarterly.*

THE ELEMENTS OF GEOMETRY.

By GEORGE CUNNINGHAM EDWARDS,

Associate Professor of Mathematics in the University of California.

16mo. Cloth. $1.10, net.

PROF. JOHN F. DOWNEY, University of Minnesota : — There is a gain in its being less formal than many of the works on this subject. The arrangement and treatment are such as to develop in the student ability to do geometrical work. The book would furnish the preparation necessary for admission to this University.

PRIN. F. O. MOWER, Oak Normal School, Napa, Cal. : — Of the fifty or more English and American editions of Geometry which I have on my shelves, I consider this one of the best, if not the best, of them all. I shall give it a trial in my next class beginning that subject.

THE MACMILLAN COMPANY,
66 FIFTH AVENUE, NEW YORK.

MATHEMATICAL TEXT-BOOKS

SUITABLE

FOR USE IN PREPARATORY SCHOOLS.

SELECTED FROM THE LISTS OF

THE MACMILLAN COMPANY, Publishers.

ARITHMETIC FOR SCHOOLS.

By J. B. LOCK,

Author of " Trigonometry for Beginners," " Elementary Trigonometry," etc.

Edited and Arranged for American Schools

By CHARLOTTE ANGAS SCOTT, D.SC.,

Head of Math. Dept., Bryn Mawr College, Pa.

16mo. Cloth. 75 cents.

"Evidently the work of a thoroughly good teacher. The elementary truth, that arithmetic is common sense, is the principle which pervades the whole book, and no process, however simple, is deemed unworthy of clear explanation. Where it seems advantageous, a rule is given after the explanation. . . . Mr. Lock's admirable 'Trigonometry' and the present work are, to our mind, models of what mathematical school books should be." — *The Literary World.*

FOR MORE ADVANCED CLASSES.

ARITHMETIC.

By CHARLES SMITH, M.A.,

Author of " Elementary Algebra," "A Treatise on Algebra,"

AND

CHARLES L. HARRINGTON, M.A.,

Head Master of Dr. J. Sach's School for Boys, New York.

16mo. Cloth. 90 cents.

A thorough and comprehensive High School Arithmetic, containing many good examples and clear, well-arranged explanations.

There are chapters on Stocks and Bonds, and on Exchange, which are of more than ordinary value, and there is also a useful collection of miscellaneous examples.

THE MACMILLAN COMPANY,
66 FIFTH AVENUE, NEW YORK.

www.ingramcontent.com/pod-product-compliance
Lightning Source LLC
Chambersburg PA
CBHW031348230426
43670CB00006B/475